Multilevel Inverters

Multilevel Inverters
Introduction and Emergent Topologies

Edited by

Ersan Kabalcı
Department of Electrical and Electronics Engineering, Faculty of Engineering and Architecture, Nevsehir Haci Bektas Veli University, Nevsehir, Turkey

Academic Press is an imprint of Elsevier
125 London Wall, London EC2Y 5AS, United Kingdom
525 B Street, Suite 1650, San Diego, CA 92101, United States
50 Hampshire Street, 5th Floor, Cambridge, MA 02139, United States
The Boulevard, Langford Lane, Kidlington, Oxford OX5 1GB, United Kingdom

Copyright © 2021 Elsevier Inc. All rights reserved.

No part of this publication may be reproduced or transmitted in any form or by any means, electronic or mechanical, including photocopying, recording, or any information storage and retrieval system, without permission in writing from the publisher. Details on how to seek permission, further information about the Publisher's permissions policies and our arrangements with organizations such as the Copyright Clearance Center and the Copyright Licensing Agency, can be found at our website: www.elsevier.com/permissions.

This book and the individual contributions contained in it are protected under copyright by the Publisher (other than as may be noted herein).

Notices

Knowledge and best practice in this field are constantly changing. As new research and experience broaden our understanding, changes in research methods, professional practices, or medical treatment may become necessary.

Practitioners and researchers must always rely on their own experience and knowledge in evaluating and using any information, methods, compounds, or experiments described herein. In using such information or methods they should be mindful of their own safety and the safety of others, including parties for whom they have a professional responsibility.

To the fullest extent of the law, neither the Publisher nor the authors, contributors, or editors, assume any liability for any injury and/or damage to persons or property as a matter of products liability, negligence or otherwise, or from any use or operation of any methods, products, instructions, or ideas contained in the material herein.

Library of Congress Cataloging-in-Publication Data
A catalog record for this book is available from the Library of Congress

British Library Cataloguing-in-Publication Data
A catalogue record for this book is available from the British Library

ISBN 978-0-12-821668-2

For information on all Academic Press publications
visit our website at https://www.elsevier.com/books-and-journals

Publisher: Brian Romer
Acquisitions Editor: Graham Nisbet
Editorial Project Manager: Sara Valentino
Production Project Manager: Sojan P. Pazhayattil
Cover Designer: Victoria Pearson

Typeset by SPi Global, India

Contents

Contributors ix
Preface xi

CHAPTER 1   **Introduction to multilevel voltage source inverters** 1
Ersan Kabalcı

  1.1 Introduction 1
  1.2 Conventional multilevel inverter topologies 6
      1.2.1 Neutral point clamped MLI.. 8
      1.2.2 Flying capacitor MLI.. 10
      1.2.3 H-bridge MLI... 11
  1.3 Soft switching and resonant multilevel inverters 15
  1.4 Fundamentals of control schemes..23
      References... 26

CHAPTER 2   **Neutral-point-clamped and T-type multilevel inverters** 29
Hasan Komurcugil and Sertac Bayhan

  2.1 Neutral-point-clamped multilevel inverters................................29
      2.1.1 Converter configuration.. 30
      2.1.2 Switching states and commutation................................... 31
      2.1.3 Modulation techniques 33
      2.1.4 Finite set model predictive control of a three-phase three-level neutral-point-clamped inverter........................ 35
  2.2 T-type inverter..43
      2.2.1 Description of T-type inverter and its operating principle 43
      2.2.2 Switch open-circuit fault 46
      2.2.3 Switch short-circuit fault.. 48
      2.2.4 Modulation of T-type inverter.. 48
      2.2.5 Influence of the switching states on DC capacitor voltages 50
      2.2.6 Simulation results 52
  2.3 Conclusion..54
      Acknowledgment 54
      References... 55

CHAPTER 3 Conventional H-bridge and recent multilevel inverter topologies 57
Ilhami Colak, Ersan Kabalcı, and Gokhan Keven

- 3.1 Introduction 57
- 3.2 H-bridge inverter topology 59
- 3.3 Common mode voltage and leakage current 60
- 3.4 Modulation strategy 63
  - 3.4.1 Bipolar SPWM 63
  - 3.4.2 Unipolar SPWM 66
  - 3.4.3 Hybrid SPWM 68
- 3.5 H5 inverter topology 70
- 3.6 H6 inverter topology 72
- 3.7 HERIC inverter 77
- 3.8 Recent H-bridge based multilevel topologies 81
  - 3.8.1 Optimized H5 topology 83
  - 3.8.2 H6-I and H6-II inverter topology 83
  - 3.8.3 H6-III 88
  - 3.8.4 H6-IV topology 88
  - 3.8.5 Passive clamped H6 topology 92
  - 3.8.6 HB-ZVR topology 98
  - 3.8.7 HBZVR-D topology 99
  - 3.8.8 Active clamped HERIC topology 99
- 3.9 Remarks and conclusion 104
- References 108

CHAPTER 4 Packed U-Cell topology: Structure, control, and challenges 111
Mohamed Trabelsi, Hamza Makhamreh, Osman Kukrer, and Hani Vahedi

- 4.1 Introduction 112
- 4.2 Packed U-cell topology 112
  - 4.2.1 Mathematical modeling 113
  - 4.2.2 Control challenges 115
- 4.3 Control techniques 115
  - 4.3.1 Finite set model predictive control 116
  - 4.3.2 Multicarrier pulse width modulation 118
  - 4.3.3 Lyapunov-based model predictive control 124
  - 4.3.4 Sliding mode control 127
  - 4.3.5 Reduced sensor control 128

4.4 Applications 135
    4.4.1 Stand-alone mode 136
    4.4.2 Grid-connected mode 136
    4.4.3 PUC5 rectifier 136
    4.4.4 PUC5-based STATCOM 137
    4.4.5 PUC5-based DVR 137
    4.4.6 PUC5 three-phase inverter 138
4.5 Commercialization challenges 141
    4.5.1 Building a mass-producible product out of a laboratory concept 142
    4.5.2 Achieving product/market requirement 142
    4.5.3 Keeping the costs/benefits/reliability balance over time 143
4.6 Conclusions 143
    Acknowledgment 144
    References 144

CHAPTER 5 Modular multilevel converters 147
Apparao Dekka, Venkata Yaramasu,
Ricardo Lizana Fuentes, and Deepak Ronanki

5.1 Introduction 148
5.2 Fundamentals of a modular multilevel converter 150
    5.2.1 Principle of operation 151
    5.2.2 Submodule configurations 152
5.3 Classical control methods 156
5.4 Pulse width modulation schemes 158
    5.4.1 Phase-shifted carrier modulation scheme 159
    5.4.2 Staircase modulation scheme 160
5.5 Submodule capacitor voltage control 161
    5.5.1 Leg voltage control 161
    5.5.2 Voltage balancing strategy 162
5.6 Current control 165
    5.6.1 Output current control 165
    5.6.2 Circulating current control 167
5.7 Applications 170
    5.7.1 HVDC transmission systems 170
    5.7.2 Offshore wind farms 172
    5.7.3 Medium-voltage motor drives 172
    5.7.4 Power quality improvement 174
5.8 Conclusions 174
    References 176

CHAPTER 6 Asymmetrical multilevel inverter topologies 181
Ilhami Colak, Ersan Kabalcı, and Gokhan Keven

- 6.1 Introduction 181
- 6.2 Asymmetric multilevel inverter with polarity generation part..184
  - 6.2.1 Multilevel DC link inverter.. 184
  - 6.2.2 Simplified asymmetric multilevel inverter 189
  - 6.2.3 Switched capacitor cell hybrid multilevel inverter......... 193
  - 6.2.4 Reduced component asymmetric multilevel inverter 195
- 6.3 Asymmetric MLI topologies without polarity generation module 197
  - 6.3.1 Asymmetric cascade multilevel inverter........................ 197
  - 6.3.2 Cascaded basic blocks multilevel inverter..................... 200
  - 6.3.3 Cascaded modified H-bridge multilevel inverter............ 203
  - 6.3.4 Cross connected sources based multilevel inverter 204
- 6.4 Remarks and conclusion 211
- References... 214

CHAPTER 7 Resonant and *Z*-source multilevel inverters 217
Oleksandr Husev and Carlos Roncero-Clemente

- 7.1 General operating principle of resonant circuits 218
- 7.2 General operating principle of impedance-source networks.....223
- 7.3 Overview of multilevel inverters...232
- 7.4 Resonant multilevel inverters: Main circuits.............................236
- 7.5 Impedance source-derived multilevel inverters: Control, benefits, and applications 239
- 7.6 Conclusions 251
- References... 252

Index 259

Contributors

Sertac Bayhan
Qatar Environment and Energy Research Institute, Hamad Bin Khalifa University, Doha, Qatar

Ilhami Colak
Department of Electrical and Electronics Engineering, Faculty of Engineering and Architecture, Nisantasi University, Istanbul, Turkey

Apparao Dekka
Department of Electrical Engineering, Lakehead University, Thunder Bay, ON, Canada

Ricardo Lizana Fuentes
Department of Environment and Energy, Universidad Católica de la Santísima Concepción, Concepcion, Chile

Oleksandr Husev
Tallinn University of Technology, Department of Mechatronics and Electrical Engineering; Department of Electrical Power Engineering and Mechatronics, TalTech University, Tallinn, Estonia

Ersan Kabalcı
Department of Electrical and Electronics Engineering, Faculty of Engineering and Architecture, Nevsehir Haci Bektas Veli University, Nevsehir, Turkey

Gokhan Keven
Department of Electronics and Automation, Vocational High School, Nevsehir Haci Bektas Veli University, Hacibektas, Nevsehir, Turkey

Hasan Komurcugil
Department of Computer Engineering, Eastern Mediterranean University, Famagusta, Mersin 10, Turkey

Osman Kukrer
Department of Electrical and Electronics Engineering, Eastern Mediterranean University, Gazimagusa, Turkey

Hamza Makhamreh
Department of Electrical and Electronics Engineering, Eastern Mediterranean University, Gazimagusa, Turkey

Deepak Ronanki
Department of Electrical Engineering, Lakehead University, Thunder Bay, ON, Canada

Carlos Roncero-Clemente
Power Electrical and Electronic Systems Research Group, University of Extremadura; Electrical, Electronic and Control Engineering Department, Extremadura University, Badajoz, Spain

Mohamed Trabelsi
Electronic and Communications Engineering, Kuwait College of Science and Technology, Kuwait

Hani Vahedi
Ossiaco Inc., Montreal, QC, Canada

Venkata Yaramasu
School of Informatics, Computing, and Cyber Systems (SICCS), Northern Arizona University, Flagstaff, AZ, United States

Preface

Multilevel inverters are one of the most widely researched power converter types in industrial and residential applications. The objective of using an inverter is to convert direct current (DC) at the input to alternating current (AC) at the output. Multilevel inverters have been proposed to overcome drawbacks of conventional two-level inverter devices, and they are being researched since a few decades. The first topologies of multilevel inverters have been implemented with developing neutral-point-clamped configuration. In the past years, multilevel inverters have gained much attention in applications of medium-voltage and high-power ranges owing to their various advantages such as low common mode voltage, decreased voltage stress on power semiconductors, low dv/dt ratio compared to two-level topologies, and low total harmonic distortion. Multilevel inverters are efficient in eliminating the harmonic component of voltage and current waveforms comparing two-level inverter topologies at the same power ratings. Improvements on microgrid and distributed generation applications such as integration of wind turbines, solar power plants, and hybrid distribution systems have promoted researches on multilevel inverters in terms of device configurations and control methods. Recent applications of multilevel inverters have a wide variety including adjustable speed drives, motor drives, active filters, integration of renewable energy sources, flexible AC transmission systems (FACTS), and static compensators. Although the most common multilevel inverter topologies are classified into three as diode clamped, flying capacitor MLI, and cascaded H-bridge, many emerging topologies and multilevel inverter configurations exist in the literature and industrial applications.

The purpose of this book is to present a broader view of emerging multilevel inverter topologies with applications and survey. The aim of the book Multilevel Inverters: *Introduction and Emergent Topologies* is to highlight emergent topologies of multilevel inverters regarding recent device topologies and control methods. A large number of specialists have joined as authors of the chapters to provide their potentially innovative solutions and research related to multilevel inverter topologies. Several theoretical researches, case analysis, and practical implementation processes are put together in this book as research and design guidelines to help graduates, postgraduates, and researchers in electrical and electronics engineering and energy systems. The book presents significant results obtained by leading professionals from the industry, research, and academic fields, which can be useful to a variety of groups in specific areas analyzed in this book.

This book has seven chapters and provides detailed introduction of each emergent multilevel inverter topologies for the readers.

Chapter 1 presents an overview of multilevel voltage-source inverters starting from conventional inverter topologies and control methods. Preliminary architectures of multilevel inverters are presented in hierarchical order. Thereafter, an introduction to resonant inverters and control methods are highlighted. Design procedures of controllers and control methods are also presented in detail in the chapter.

Chapter 2 introduces the neutral-point-clamped (NPC) and T-type multilevel inverters. Operating principles, switching states, switch fault analysis, influence of switching states on dc capacitor voltages, and modulation of NPC and T-type inverter topologies are discussed in the subsections. This chapter includes simulation studies presented for NPC and T-type inverters.

Chapter 3 is a comprehensive introduction of conventional H-bridge multilevel inverters and enhanced H5 and H6 topologies with control methods and simulation analyses. In addition to device topologies, modulation and control methods are presented. Comprehensive analysis and modeling studies are highlighted in this chapter.

Chapter 4 introduces packed U-cell topology as one of the emergent multilevel inverters. Control methods, simulation studies, and applications are presented in detail. A comprehensive study on the different types of packed U-cell configurations, modeling, control techniques, applications, and commercialization challenges is given in the chapter.

Chapter 5 highlights a comprehensive study on the operation, control, and application of modular multilevel converters. The operation and features of the most widely used submodules including H-bridge, full bridge, flying capacitor, and cascaded half-bridge topologies are discussed. Finally, the commercial applications of MMC are briefly summarized, which proves the importance and potentiality of MMC for high-power industrial applications.

Chapter 6 deals with introducing asymmetrical multilevel inverter topologies, where simplified switched capacitors, cascaded asymmetrical, and similar other applications are presented in detail. In addition to regular topologies, novel asymmetrical multilevel inverters are introduced in the subheadings of the chapter.

Chapter 7 presents the resonant and impedance-source multilevel inverters. The general operation principle of the resonant circuits is introduced and a soft switching operation is presented. Finally, a combination of resonant circuits and impedance networks with multilevel inverters are discussed in the chapter.

Multilevel Inverters': *Introduction and Emergent Topologies* aims to help electrical and electronics engineers, who intend to conduct research in the field of power electronics and converters. The book also explores the recent progress on several application areas of multilevel inverters including machine drives, active filters, static var. compensator, microgrid control and their performance evaluation in terms of grid integration. I hope that this book will be helpful for young researchers and practitioners in the area of electrical engineering. Editor and authors have made every effort to make this book a useful and comprehensive one for its readers.

CHAPTER 1

Introduction to multilevel voltage source inverters

Ersan Kabalcı

Department of Electrical and Electronics Engineering, Faculty of Engineering and Architecture, Nevsehir Haci Bektas Veli University, Nevsehir, Turkey

1.1 Introduction

DC-AC converter devices, also known as inverters, are used to convert DC supply at their input to an AC waveform at the output. This converter architecture is commonly known as an inverter, which is also used as a general term in this Chapter. The DC supply type determines the definition of the inverter as a current source inverter (CSI) or voltage source inverter (VSI), depending on whether it is fed by a current source or voltage source, respectively. A typical CSI comprises series inductors connected to a DC supply source, while VSI topology has parallel capacitors connected to the DC supply bus. The inverter type also determines the controlled waveform at the output, whether a CSI to control current source or a VSI for controlling the voltage source. Since the VSI covers the vast majority of inverters used in industrial, residential, and other many areas, this book and related chapters focus on VSI topologies instead of CSI. However, the device topologies and control methods are similar in many aspects [1, 2].

Inverters have found a rapid evaluation in industrial applications compared to other power switching devices, due to their wide range of use. The foremost applications of inverters include adjustable speed drives (ASDs) for AC motors, induction heating systems, uninterruptible power supplies (UPSs), AC power supplies from several DC sources, traction control drives, and recent vector-controlled drives in industrial applications. In addition to industrial applications, inverters are widely used in generation, transmission, distribution, and renewable energy source integration areas where power conditioning plays a crucial role. In this regard, flexible AC transmission systems (FACTS), VAR compensators, static VAR compensators (STATCOMs), distribution STATCOM (D-STATCOM) active harmonic filters (AHFs), unified power flow controllers (UPFCs), and grid interactive inverters used to integrate renewable energy sources such as wind turbines, solar power plants, fuel cells, and biogas plants can be listed as power system applications of inverters. Although almost all application areas require characteristic solutions, the device topology and control methods used to operate inverters are similar across applications. The device topologies can be classified into two main categories: two-level topologies that are built with conventional six-pulse converter structure in three phase, and multilevel topologies that have been improved for many years in order to provide several advantages against two-level topology. The multilevel inverters (MLIs) have received attention due to their increased output voltage level, as the name implies, leading to decreased voltage stress on switching devices, lower dv/dt ratios, lower common mode voltage, and decreased electromagnetic interference (EMI) noise and total harmonic distortion (THD) ratios as compared to two-level topology [3–5].

The conventional two-level inverters were the industry standard for many years, until the late 1980s. The improvements in power semiconductors, such as the insulated gate bipolar transistor (IGBT) and metal oxide semiconductor field effect transistor (MOSFET), and reduced costs have leveraged research on MLI topologies. One of the first research projects in this area was proposed by Nabae et al. in [6], in which improvements in neutral point clamped (NPC) inverter topology were presented. The NPC topology proposed for increasing motor drive efficiency by decreasing the harmonic content has been accepted as the first application of MLIs. It has proven that it is possible to decrease THD ratios in the output voltage and current waveforms, and the result is improved efficiency of the inverter. Moreover, the proposed topology was suitable for control at higher switching frequencies and operation with several pulse width modulation (PWM) schemes. The improvements in inverters can be dealt with in two areas, as device topologies and control methods, and they are also presented in more detail in later chapters of this book.

The NPC topology has triggered enhancement of other MLI topologies, with the most widely known being described as flying capacitor and H-bridge structures, as seen in Fig. 1.1B and C, respectively. The output voltage waveform and frequency of a VSI are controlled by using PWM methods, which provide amplitude and frequency adjustments. The regular PWM method is based on comparison of a modulating sinusoidal waveform to a carrier triangular waveform. The main objective of the sinusoidal signal is to determine the line frequency of the output voltage, while the switching frequency is adjusted by the carrier signal. The comparison generates pulse signals with different widths that denote the definition of PWM theory. The fundamental block diagram of an ASD is illustrated in Fig. 1.2, where the grid voltage has been rectified by a basic uncontrolled rectifier and supplied to the inverter for generating variable voltage and variable frequency output to drive any AC motor. The ASD applications are based on the varying sinusoidal waveform frequency in the modulating signal, which is typically changed from 50 to 600 Hz in industrial variable frequency drives (VFDs), while the frequency of the carrier signal can be increased up to 20 kHz [1–3].

The ASD block diagram shown is based on single directional operation due to the uncontrolled rectifier, where the power flow is provided from grid to motor. In this general scheme, it is not possible to recover braking or deceleration energy induced in the motor. However, recent applications of electric vehicles and wind turbines are based on bidirectional ASD operation in which the uncontrolled rectifiers are replaced with IGBT or MOSFET inverters. Thus both rectifier and inverter can be

1.1 Introduction

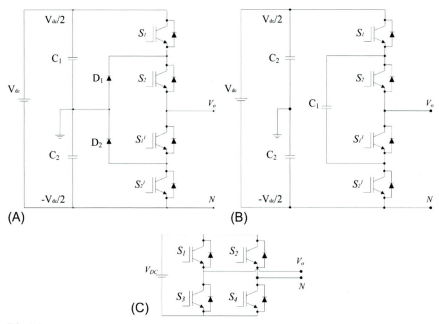

FIG. 1.1

Fundamental three-level MLI topologies: (A) neutral point clamped MLI, (B) flying capacitor MLI, (C) H-bridge MLI.

used to shift the operation due to control methods and switching angles. This improved ASD architecture enables the motor to be operated in a four-quadrant scheme with regenerative braking and generator modes.

MLI topologies can be classified according to application areas, conversion strategies, performance parameters, and so on. In this chapter, MLIs are classified according to supply types and technological enhancements as conventional, resonant, and reduced number topologies, as shown in Fig. 1.3. The NPC and flying capacitor (FC)

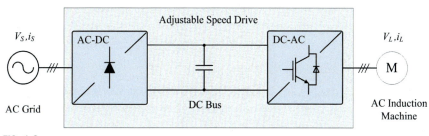

FIG. 1.2

Block diagram of an ASD.

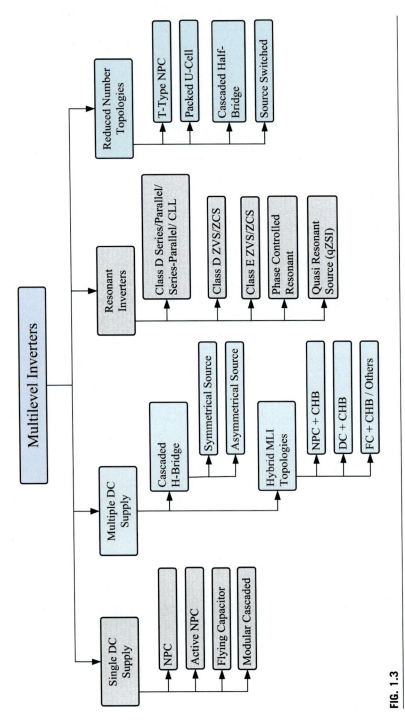

FIG. 1.3
General list of MLI topologies based on supply and enhanced topology types.

topologies are widely known single-supply inverters, while cascaded H-bridge (CHB) topologies require multiple supply sources. The ratio of supply sources determines the type of cascaded H-bridge topology as based on either a symmetrical or asymmetrical source, as seen in the second column of Fig. 1.3. In addition to source-based structure, cascaded topologies are improved via hybrid structures, such as a combination of NPC and CHB, diode clamped (DC) and CHB, FC and CHB, and similar other approaches. These topologies are improved in order to use the advantages of different topologies in a single inverter structure. Novel MLI topologies have been proposed in resonant network configurations that require an increased number of switching devices and passive components, causing increased cost and complexity. Although the high number of switching semiconductors increases switching losses due to the physical state of regular silicon devices, the resonant MLI topologies tackle this issue with zero voltage or zero current switching methods. On the other hand, these topologies provide appropriate solutions for decreasing leakage current and THD ratios while increasing overall efficiency and power quality due to the resonant network and control methods [7].

Quasi-resonant source inverters (qZSI) provide the opportunity of using CSI and VSI selection in a single inverter. The LC network located between the DC bus and switching devices can be controlled to operate the inverter either in CSI mode or VSI mode. Another emerging group of MLI topologies is known as reduced number devices, as seen on the right side of Fig. 1.3. The reduced number topology is aimed at overcoming the increased size and cost of conventional topologies, as increased numbers of switches cause complexity in controller design and increase the required measurement nodes.

Although it is possible to find a wide variety of MLI topologies in the literature, the most widely used and accepted industrial topologies are presented in this section. The T-type NPC topologies are assumed as an industry standard in many drivers and conversion infrastructures as an alternative to conventional NPC topology. On the other hand, the packed U-cell (PUC) topology is another alternative to decrease switch numbers and complexity of the controller. These topologies are introduced briefly in the following sections of this chapter, and in detail in the following chapters of the book.

Inverter control methods are based on modulation schemes that play a crucial role in the overall efficiency of power conversion. The modulation methods determine the power factor, THD ratios, and voltage and frequency rates of output voltage. The MLI staircase output waveform is generated by the multiple number of switching devices in the MLI topology, and it is dependent on control of the modulator, which generates switching signals in variable duty cycle rates. The switching methods are described as fundamental frequency switching, high frequency switching, and variable frequency switching. Fundamental frequency switching operates in the same way as line frequency commutation, where each switch is triggered once or twice in a single cycle. High-frequency switching is performed by a carrier-based PWM scheme in which the carrier frequency reaches several kilohertz as usual [8, 9]. The widely used high-frequency switching methods are addressed in the

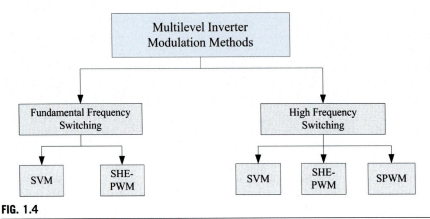

FIG. 1.4

MLI modulation methods.

literature as sinusoidal PWM (SPWM), selective harmonic elimination PWM (SHE-PWM), space vector PWM (SVM), and third harmonic injection PWM (THI-PWM). All of these modulation schemes can be improved by single- or multicarrier signals in order to increase efficiency and harmonic elimination features.

The SVM and SHE-PWM methods are also convenient to be used in fundamental frequency switching, among others. On the other hand, the SVM scheme finds widespread use in ASD drives in two-level and three-level inverters as an industry standard [4].

This chapter gives an introduction to conventional and recent MLI topologies, with a featured section presented on reduced number device topologies. On the other hand, fundamentals of MLI control schemes have been drawn based on the schemes depicted in Fig. 1.4, while applications of MLIs are dealt with in the last subsection. This chapter is basically an introduction to the content of this book, with MLI topologies, control methods, and application areas presented in detail.

1.2 Conventional multilevel inverter topologies

The multilevel inverter topologies have been derived from simple single-phase devices configured as half-bridge and full-bridge. High-power applications are performed by using MLI configurations of these low-power single-phase topologies, which are used as AC power supplies in industry. The half-bridge VSI topology is illustrated in Fig. 1.5A, while Fig. 1.5B shows the full-bridge topology in single-phase configuration. The neutral point of these devices is generated by two large capacitors connected in parallel to the source, where the coupling point of the capacitors is used as the neutral point. Each capacitor provides half of the supply voltage, and the large capacity of the capacitors eliminates the low ordered harmonics in the operation. The control methods applied in both VSI topologies should prevent a short circuit of the power source V_{DC} by switching each device in the same

1.2 Conventional multilevel inverter topologies

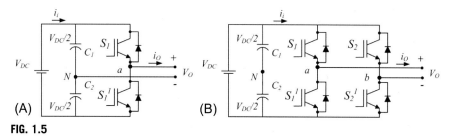

FIG. 1.5

Single-phase VSI topologies: (A) half-bridge, (B) full-bridge.

leg at different time intervals. The operation of a half-bridge comprises two predefined and a single undefined switching state. The predefined states are used to turn on S_1 while in the complementary one S_1^1 is turned off; the second state inverts this order. In the first state, the output voltage is obtained at the level $V_{DC}/2$ while the second state generates $-V_{DC}/2$ output voltage. The undefined state turns both switches off and causes generation of either $V_{DC}/2$ or $-V_{DC}/2$ voltage levels, according to conduction of D_1 or D_1^1 diodes.

The full-bridge VSI presents the same operation as the half-bridge, where the second phase arm is added to the configuration for generating a neutral point on the load side. The operation of the switching order is controlled to prevent a short circuit of the supply voltage by turning on the S_1 while S_1^1 is turned off or a similar operation is applied for S_2 and S_2^1. The predefined switching orders are increased to four in this topology, while one undefined switching state remains. The switching orders and conducting components are listed in Table 1.1. In order to ensure AC output voltage is supplied in any switching state, the undefined state seen in the fifth order should be avoided. Short-circuit protection is also ensured by performing complementary switching for devices in the same phase arm. The modulation scheme should be designed and implemented considering these criteria [3–5].

Table 1.1 Switching states of single-phase full-bridge VSI.

Order	Switching state	V_{aN}	V_{bN}	V_o	Conducting device
1	S_1 and S_2^1 on	$V_{DC}/2$	$-V_{DC}/2$	V_{DC}	S_1 and S_2^1 if $i_o > 0$ D_1 and D_2^1 if $i_o < 0$
2	S_1^1 and S_2 on	$-V_{DC}/2$	$V_{DC}/2$	$-V_{DC}$	D_1^1 and D_2 if $i_o > 0$ S_1^1 and S_2 if $i_o < 0$
3	S_1 and S_2 on	$V_{DC}/2$	$V_{DC}/2$	0	S_1 and D_2 if $i_o > 0$ D_1 and S_2 if $i_o < 0$
4	S_1^1 and S_2^1 on	$-V_{DC}/2$	$-V_{DC}/2$	0	D_1^1 and S_2^1 if $i_o > 0$ S_1^1 and D_2^1 if $i_o < 0$
5	All switches are off	$-V_{DC}/2\ V_{DC}/2$	$V_{DC}/2\ -V_{DC}/2$	$-V_{DC}\ V_{DC}$	D_1^1 and D_2 if $i_o > 0$ D_1 and D_2^1 if $i_o < 0$

Although many modulation schemes have been proposed in the literature, the PWM-based methods are essential for VSI control. Modulation signals, generated switching pulses, and output voltages of half-bridge and full-bridge VSIs are shown in Fig. 1.6A and B, respectively. The modulating sinusoidal waveform is compared to a triangular carrier signal, as seen in the first axes of the figures. The sinusoidal waveform determines the frequency of the line voltage, while the frequency of the carrier adjusts the frequency of pulses applied to switching devices. The amplitude ratio of the modulating signal to the carrier is defined as the modulation index m_i and the ratio of carrier frequency to modulation signal frequency is normalized frequency m_f [3, 4]. The modulation techniques and equations are presented in the last section of this chapter.

The last axes of Fig. 1.6 show the output voltage ratios of the half-bridge and full-bridge, where the values are calculated using the following equations, respectively:

$$V_O = V_{aN} = \frac{V_{DC}}{2} m_i \tag{1.1}$$

$$V_O = V_{ab} = V_{DC} m_i \tag{1.2}$$

The equations prove that the output of the half-bridge VSI is a ratio of $V_{DC}/2$ while the full-bridge output is twice that of the half-bridge due to the second phase arm in the configuration. On the other hand, the output voltage is dependent on the modulation index ratio, since it provides gain at the output voltage. The operation modes of the modulator can be adjusted as in the linear region where $m_i < 1.0$, in the overmodulation region where $1.0 < m_i < 3.24$, or in square-wave operation where $m_i > 3.24$ ratio. The linearity between modulation index and output voltage is only obtained in the linear operation region, while in the overmodulation region the output voltage increases but low-order harmonics are generated [3].

The single-phase inverters are suitable in low-power applications such as residential or small-load plants. When medium- or high-power applications are required, three-phase VSI topologies are essential to extend the operation range of the inverters. The following sections describe the most widely known VSI topologies in three-phase configurations.

1.2.1 Neutral point clamped MLI

The NPC, which is also called a diode clamped MLI topology, was proposed by Nabae et al. as the first three-level MLI in 1981 [6]. The topology is supposed to be an industrial driver dealing with harmonic contents and increasing the overall system efficiency. The proposed three-level three-phase NPC inverter is shown in Fig. 1.7. The neutral point comprises the common coupling of two large capacitors at the input of the inverter. The coupling points of diodes at each phase leg are also connected to the neutral point of the circuit and each phase leg comprises four switches with antiparallel diodes in each one.

The characteristic features of NPC topology are as follows: it provides a third output level (0 V_{DC}) in addition to $V_{DC}/2$ and $-V_{DC}/2$, and the output waveforms present

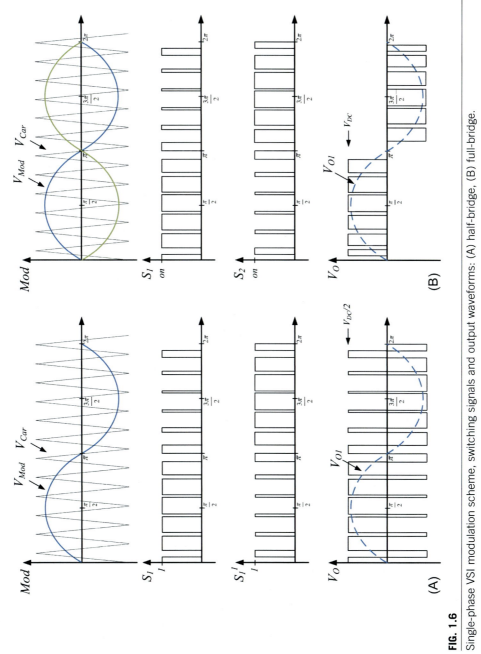

FIG. 1.6
Single-phase VSI modulation scheme, switching signals and output waveforms: (A) half-bridge, (B) full-bridge.

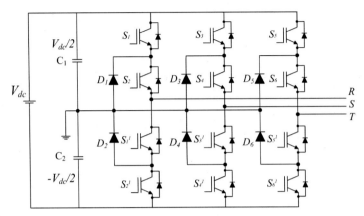

FIG. 1.7

Three-phase three-level conventional NPC MLI topology.

decreased harmonic content compared to conventional two-level topologies. The key devices of this topology are clamping diodes, which are essential to generate the third voltage level differing from the two-level inverter. The clamping diodes require different voltage ratings for reverse voltage-blocking due to the fact that each triggered switch is only required to block a voltage level of $V_{DC}/(m-1)$, where m denotes the level of output waveform. Since each blocking diode voltage is rated at the same value as the active device voltage rating, the number of diodes required for each phase can be calculated as $(m-1)(m-2)$. The following equations are used to determine the required device numbers to form a given level of a diode clamped MLI. If m is assumed as the number of levels, the number of capacitors at the DC side C_{DC} is defined by Eq. (1.3). The number of freewheeling diodes (D_{FW}) per phase and the number of clamping diodes (D_c) can be calculated using Eqs. (1.4) and (1.5), respectively [4, 10].

$$C_{DC} = m - 1 \tag{1.3}$$

$$D_{FW} = 2(m - 1) \tag{1.4}$$

$$D_C = (m - 1)(m - 2) \tag{1.5}$$

The power range and operating voltage of an NPC topology can be extended to higher values by adding several switching devices at each phase leg and increasing the clamping diodes at the input of the switching block, which is the main reason for defining this topology as a clamping diode MLI.

1.2.2 Flying capacitor MLI

The flying capacitor topology was proposed to overcome unbalanced voltage problems and increased number of diodes in the DC MLI topology. The device configuration of a three-phase three-level FC MLI is shown in Fig. 1.8, where each phase

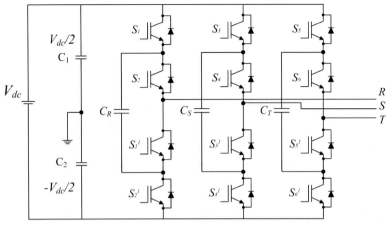

FIG. 1.8

Three-phase three-level conventional FC MLI topology.

leg is identically composed. The voltage level of the FC topology is similar to the NPC, where an *m*-level converter generates (2*m*–1) level line voltages. The dc-bus of this topology comprises (*m*–1) capacitors, where the two adjoining capacitor legs give the size of the voltage steps in the output waveform. In an *m*-level structure, FC-MLIs require (*m*–1)(*m*–2)/2 auxiliary capacitors per phase compared to the DC-MLI topology. The device configuration of a three-phase three-level FC MLI topology can be extended to a five-level configuration, as seen in Fig. 1.9; the number of switching devices will be two times that of the three-level topology and the flying capacitors will be increased to six instead of two. The switching devices are complementarily triggered in the order of (S_{A1}-S_{A1}^1), (S_{A2}-S_{A2}^1), (S_{A3}-S_{A3}^1), and (S_{A4}-S_{A4}^1). The dc-bus capacitor C is used as the energy storage device, while C_{A1}, C_{A2}, C_{A3} are the flying capacitors. The switching states of one phase leg in a five-level FC MLI are shown in Table 1.2 [4, 11–13].

The table presents four alternative states for generating $V_{DC}/4$ and $-V_{DC}/4$ levels, while the zero voltage level can be generated in six different switching states. The charging and discharging states of the flying capacitors are also listed in Table 1.2. The switching states given for the positive half-cycles will invert the capacitor states in negative half-cycles, but "no change ≈" situations remain the same in both half-cycles.

The FC MLI topology provides redundancy of switching-state variations, a single capacitor control device, equal voltage stress on devices, and alternative switching states based on current rating of capacitors [11].

1.2.3 H-bridge MLI

Despite the aforementioned topologies, the H-bridge MLI configuration comprises four switching devices and there is no clamping diode or flying capacitor requirement for the device. Another difference from other topologies is the separate DC

12 CHAPTER 1 Introduction to multilevel voltage source inverters

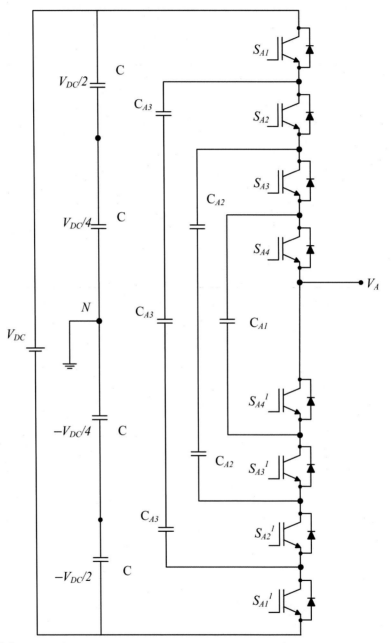

FIG. 1.9
Single-phase leg of a five-level conventional FC MLI topology.

Table 1.2 Switching states of single-phase full-bridge VSI.

Output voltage level	Switching state S_{A1}	S_{A2}	S_{A3}	S_{A4}	C_{A3}	C_{A2}	C_{A1}
$V_O = V_{DC}/2$	on	on	on	on	≈	≈	≈
$V_O = V_{DC}/4$	on	on	on	off	≈	≈	+
	on	on	off	on	≈	+	−
	on	off	on	on	+	−	≈
	off	on	on	on	−	≈	≈
$V_O = 0$	off	off	on	on	≈	−	≈
	off	on	off	on	−	+	−
	off	on	on	off	−	≈	+
	on	off	off	on	+	≈	−
	on	off	on	off	+	−	+
	on	on	off	off	≈	+	≈
$V_O = -V_{DC}/4$	on	off	off	off	+	≈	≈
	off	on	off	off	−	+	≈
	off	off	on	off	≈	−	+
	off	off	off	on	≈	≈	−
$V_O = -V_{DC}/4$	off	off	off	off	≈	≈	≈

≈: no change in the state, +: charging, −: discharging.

source requirement in the H-bridge. In a multiphase application, the H-bridge topology requires separate supply sources for each inverter cell. The cascaded connection of cells provide higher output levels with a rational increment in switching devices, and it is composed of the least number of active devices in the cascaded H-bridge (CHB) MLI topology. The CHB MLI topology is defined as symmetrical when it is configured with identical DC voltage sources at each cell, while the asymmetrical topology is obtained by using rational DC source levels [3, 14–17]. Fig. 1.10 illustrates a three-level H-bridge MLI application in three-phase, where each phase cell is fed by separate DC sources. The output voltage levels for each cell are obtained as shown in Eq. (1.6) according to switching orders and states of active devices in the topology.

$$V_X = \begin{cases} +V_{DC} & S_{x1}, S_{x2}^1 \text{ on} \\ 0 & S_{x1}, S_{x2} \text{ on} \\ -V_{DC} & S_{x1}^1, S_{x2} \text{ on} \end{cases} \quad (1.6)$$

In order to increase the number of voltage levels from three to five or more, each phase leg should be connected in serial with new H-bridge cells to comprise the CHB topology. If the number of the DC source is denoted with k, the output voltage levels per phase m and the line voltage levels l are calculated as shown in Eqs. (1.7) and (1.8), respectively.

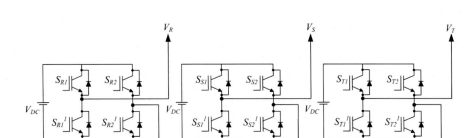

FIG. 1.10

Three-phase three-level H-bridge MLI topology.

$$m = 2k + 1 \tag{1.7}$$

$$l = 2m - 1 \tag{1.8}$$

A single phase leg of a CHB MLI is represented in Fig. 1.11, where the DC voltage levels would be the same or at the ratios of $V_{DC}{:}n{\cdot}V_{DC}$, where n may be 2 or 3 based on a binary or trinary asymmetrical topology configuration. The resultant output voltage levels of asymmetrical topologies are also calculated using Eqs. (1.7) and (1.8), where k is obtained based on $(n{\cdot}V_{DC})$ value of DC sources.

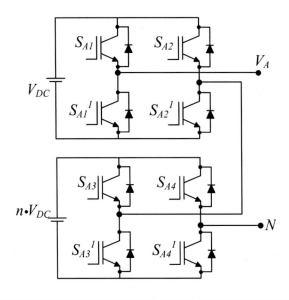

FIG. 1.11

Single phase leg of a CHB MLI topology.

The symmetrical CHB MLI topology given in Fig. 1.11 generates a five-level output voltage with eight switching devices, while the binary asymmetrical CHB generates seven-level and the trinary asymmetrical topology generates nine-level outputs in the same configuration. It can be stated that the main contribution of the CHB MLI topology is its capability to increase output voltage levels by increasing the $V_{DC}:n \cdot V_{DC}$ ratio. Moreover, the cell numbers in the cascaded configuration can also be increased to obtain a higher number of output levels [10, 14, 17, 18].

1.3 Soft switching and resonant multilevel inverters

A great deal of research has been done to reduce switching losses caused by PWM methods used for the control of inverters. Hard-switching methods that do not track the load side and turn the switching devices on and off in a predefined order generate serious power losses during switching times. Soft-switching methods have been proposed not only to reduce the switching losses but also to decrease EMI noise. The switching devices S+ and S− shown in a phase leg of an inverter in Fig. 1.12A are respectively turned on and turned off. If it is assumed that the load current is fixed at the value I_0 at each switching interval due to inductance occurring in the circuit, the switching transitions of the S− switch are obtained as shown in Fig. 1.12B. It is

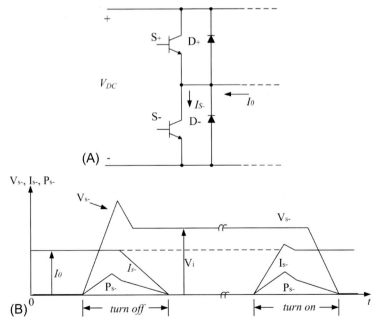

FIG. 1.12

Zero transition switching: (A) single phase leg; (B) switching losses.

assumed that the I_O currents pass through the S− switch as I_{S-} current. In this case, a turn-off signal is applied to the S− switch, resulting in switching the voltage to excess input source voltage V_i and then returning to the V_i value. The current flowing through the S− switch will decrease to zero at the end of the fall time, which enables the D+ diode to carry the current and causes a switching power loss P_{s-} in this interval. The operation is reversed in the rise time of the switch current after the turn-on signal, which causes the I_{S-} current to rise to the I_O value.

In the presented hard-switching scheme, a switching power loss occurs at P_{s-} due to the intersection of the switch current and voltage values during turn-off and turn-on intervals, as seen in Fig. 1.12B [3, 19, 20]. The average value of switching power loss P_T occurring in a switching device is proportional to the switching frequency applied to the device. The switching losses can be minimized by applying the turn-on and turn-off signals at the zero voltage or zero current states in the switching inverters. In ideal cases, the current and voltage values should be zero during switching transitions. A comparison of switching power losses occurring in any switching device during hard-switching, snubbered control, and soft-switching control is illustrated in Fig. 1.13. Despite the hard-switching schemes, dissipative snubber networks can decrease the switching power losses by decreasing dv/dt or di/dt of the power devices. However, this does not provide a higher decrement on power losses as seen in the soft-switching methods, shown in the figure. It is noted in [3] that a

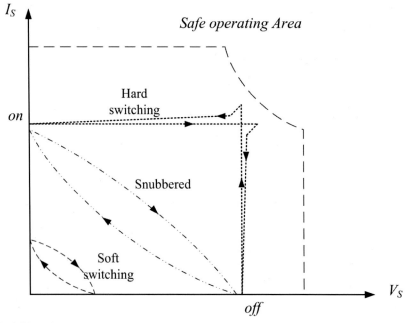

FIG. 1.13

Switching methods applied to power semiconductors.

great deal of research was performed on zero voltage switching (ZVS) and zero current switching (ZCS) methods in the context of soft-switching developments in the 1980s.

One of these topologies improved to minimize switching power losses is known as a resonant inverter. The resonant switching devices are improved to be operated at the resonance intervals of inductive and capacitive output filters as well as in resonant converters. The resonance networks can be improved in series, as seen in Fig. 1.14, with the frequency spectrum in parallel, or in series-parallel in order to generate a frequency response. The L and C create a resonant frequency, while R is used to control the resonance curve. The most important issue in the resonant frequency f_r is determined by reactive elements, and it denotes the frequency at which released energy of any reactive element is stored by another reactive element at this frequency or around this frequency.

The total power seen in a resonance network is roughly equal to the dissipated power in the ohmic elements, and thus the dissipated or stored energy of the system would be at the maximum resonance value. The limit of bandwidth seen in Fig. 1.14 depends on and varies with inductor and capacitor values in the resonant network. The slope angle of the curve is rational with self-resistance of the inductor, and less self-resistance yields higher inductance quality (Q). The value of Q is calculated with the ratio of inductive reactance to self-resistance of an inductor.

The resonant inverters are also composed of switching devices such as IGBT or MOS controlled devices, where the output of the inverter includes a resonance network that enables ZCS or ZVS control. The resonant inverters are represented in Fig. 1.15, where the most widely used ones belong to Class D topologies as seen in the figure. Each configuration can be controlled with ZVS or ZCS methods, as listed under Class E topologies. The main topologies are listed as current source or voltage source as in regular inverter devices. The voltage source resonant inverters

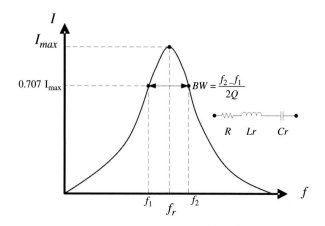

FIG. 1.14

Frequency spectrum of series resonant network.

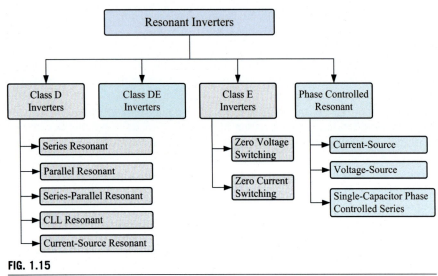

FIG. 1.15

Resonant inverter topologies and device types.

create less voltage stress on switching devices but require higher inductor and capacitor values at the filter section [3, 19, 20].

The Class D resonant inverters use a series resonance network at the output of the inverter and a high-enough quality factor (Q) will ensure the inverter current is sinusoidal and switching device currents are half-wave sinusoidal. The most important advantage of Class D inverters is their suitable use in high-voltage applications due to low voltage stress on switches. The common topologies of Class D resonant inverters are illustrated in Fig. 1.16 for half-bridge and full-bridge configurations. The half-bridge series resonant inverter, which is composed of bidirectional S_1 and S_2 switches and a series LC resonance network, is seen in Fig. 1.16A. The full-bridge version of a series Class D resonant inverter is seen in Fig. 1.16B, with four switches and diodes providing bidirectional power flow. The parallel Class D resonant inverters are shown in Fig. 1.16C and D for half-bridge and full-bridge topologies, respectively.

In this parallel configuration, a blocking capacitor Cc is connected to the output of the inverter in addition to resonance inductance Lr and resonance capacitor Cr, and the load resistance is series connected to this capacitor. The half- and full-bridge series-parallel Class D topologies seen in Fig. 1.16E and F are similar to parallel resonant inverters but require a capacitor series connected to a resonance inductor. On the other hand, it draws a similar configuration to a series topology due to its parallel capacitor to load resistance. In both cases, the presented topology is similar to series and parallel resonant inverters. The LCL resonant inverter topology shown in Fig. 1.16G is composed of an L_1-C-L_2 resonance network at the output of a half-bridge inverter. The resonance capacitor is series connected to center-tapped

1.3 Soft switching and resonant multilevel inverters

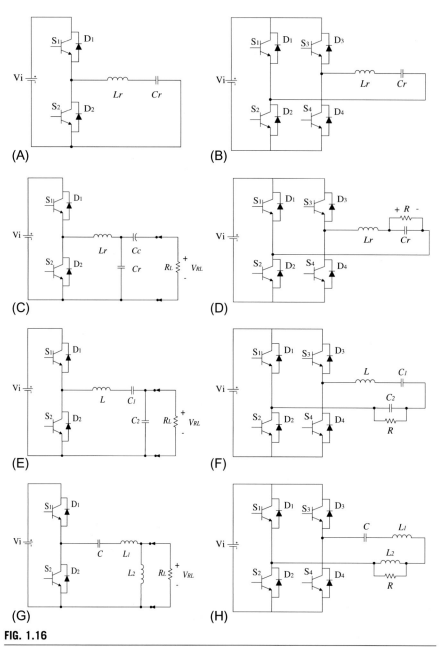

FIG. 1.16

Class D resonant inverter topologies: (A) half-bridge series, (B) full-bridge series, (C) half-bridge parallel, (D) full-bridge parallel, (E) half-bridge series-parallel, (F) full-bridge series-parallel, (G) half-bridge LCL, (H) full-bridge LCL.

inductors L_1-L_2, and the load is parallel connected to inductor L_2. The full-bridge alternative of this LCL resonant inverter is shown in Fig. 1.16H. The Class D LCL resonant inverter is presented in detail as a design example in the following part of this section.

The full-bridge LCL resonant inverter topology is given in Fig. 1.17, where the input voltage v_{in} is the square wave of magnitude $2V_I$ as follows:

$$v_{in} = \begin{cases} V_I, & 0 < \omega t \leq \pi \\ -V_I, & \pi < \omega t \leq 2\pi \end{cases} \tag{1.9}$$

The fundamental component of v_{in} is calculated based on the $v_{i1} = V_m \sin(\omega t)$ equation where

$$V_m = \frac{4}{\pi} V_I = 1{,}273 V_I \tag{1.10}$$

Thus the rms value of V_m is obtained as 0.9. The inductance, equivalent inductances, and resonance frequency are calculated as follows:

$$A = \frac{L_1}{L_2},\; L = L_1 + L_2 = L_2(1+A) = L_1\left(1 + \frac{1}{A}\right) \tag{1.11}$$

$$\omega_0 = \frac{1}{\sqrt{LC}} = \frac{1}{\sqrt{(L_1 + L_2)C}} \tag{1.12}$$

The characteristic impedance Z_0 and load quality factor Q_L are determined by the following equations:

$$Z_0 = \omega_0 L = \frac{1}{\omega_0 C} = \sqrt{\frac{L}{C}} \tag{1.13}$$

$$Q_L = \frac{R_i}{\omega_0 L} = \omega_0 C R = \frac{R_i}{Z_0} \tag{1.14}$$

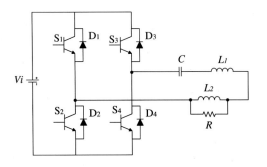

FIG. 1.17

Class D full-bridge LCL resonant inverter.

1.3 Soft switching and resonant multilevel inverters

The boundaries of an LCL inverter in capacitive and inductive operation are determined according to the resonance frequency. In this regard, the load quality factor at the resonance frequency is obtained as shown in Eq. (1.15):

$$Q_r = \frac{1}{\omega_0 R_s C} = \frac{\omega_r (L_1 + L_2)}{R_s} \quad (1.15)$$

On the other hand, the input impedance of a resonance network is calculated depending on inductance ratio, quality factor, and resonance frequency as follows:

$$Z = \frac{R_i \left\{ (1+A) \left[1 - \left(\frac{\omega_o}{\omega}\right)^2\right] + j\frac{1}{Q_L}\left(\frac{\omega}{\omega_o}\frac{A}{A+1} - \frac{\omega_o}{\omega}\right) \right\}}{1 - jQ_L\left(\frac{\omega_o}{\omega}\right)(1+A)} = |Z|e^{j\psi} \quad (1.16)$$

$$Z = R_s + jX_s$$

The current passing through capacitor C is defined by $i = I_m \sin(\omega t - \psi)$ where I_m is dependent on aforementioned parameters and the voltage transfer function of the resonant network M_{vr} [20]:

$$I_m = \frac{4 V_i M_{vr}}{\pi Z_o Q_L} \sqrt{1 + \left[Q_L\left(\frac{\omega_o}{\omega}\right)(1+A)\right]^2} \quad (1.17)$$

The output power of inverter P_{Ri} is expressed in Eq. (1.18) while the conduction power loss P_r is calculated using Eq. (1.19):

$$P_{Ri} = \frac{V_{Ri}^2}{R_i} = \frac{8 V_I^2}{\pi^2 Z_o Q_L \left\{ (1+A)^2 \left[1 - \left(\frac{\omega_o}{\omega}\right)^2\right]^2 + \frac{1}{Q_L^2}\left(\frac{\omega}{\omega_o}\frac{A}{A+1} - \frac{\omega_o}{\omega}\right)^2 \right\}} \quad (1.18)$$

$$P_r = \frac{r I_m^2}{2} = \frac{8 r V_I^2 M_{vr}^2 \left\{ 1 + \left[Q_L\left(\frac{\omega_o}{\omega}\right)(1+A)\right]^2 \right\}}{\pi^2 Z_o^2 Q_L^2} \quad (1.19)$$

If the switching losses are neglected, the efficiency of a full-bridge LLC resonant inverter is obtained as follows:

$$\eta_I = \frac{P_{Ri}}{P_1} = \frac{P_{Ri}}{P_{Ri} + P_r} = \frac{1}{1 + \frac{r}{R_i}\left\{1 + \left[\frac{R_i \omega_o}{Z_o \omega}(1+A)\right]^2\right\}} \quad (1.20)$$

The recent applications of resonant power converters are inherited from the conventional H-bridge topology and resonance network composed of a combination of L-C elements. The charge and discharge cycles of the resonance network are controlled for ensuring the maximum energy transfer to the load by minimizing the switching losses. Such resonant inverters are improved by use of a high-frequency (HF) transformer, which provides galvanic isolation on the load side. The use of a HF transformer enables increasing the switching frequency, which is managed by soft switching methods to decrease leakage inductance and EMI noise. Fig. 1.18

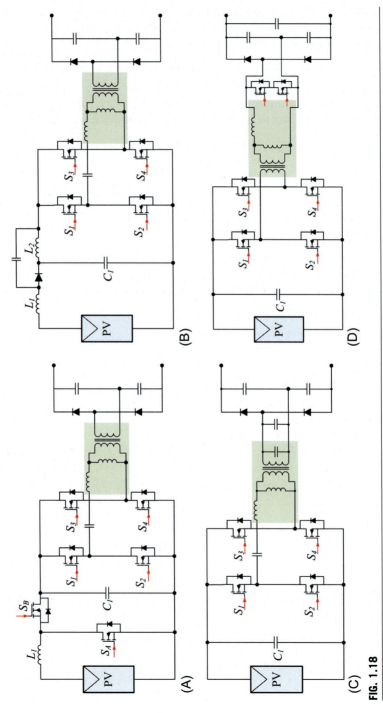

FIG. 1.18

The resonant microinverter applications: (A) two-stage topology with synchronous series-resonant converter, (B) quasi-Z source series resonant converter, (C) LLCC series-parallel resonant converter, (D) series resonant converter with bidirectional switch [7].

illustrates some emerging inverter topologies that are used in interfacing solar modules as renewable energy sources. The application circuits are generally identified as converters in these applications where single or two-stage inverter configurations are utilized. The resonant network can be in series, parallel, or series-parallel connection, as seen in the figures.

Fig. 1.18A shows a synchronous series resonant inverter operated in boost mode and followed by a series resonant network composed of an HF transformer. This topology represents increased efficiency based on ZVS or ZCS control decreasing the switching losses. On the other hand, two-way power flow of the inverter section ensures reliable operation of the HF transformer by discharging the primary side of the transformer. The quasi-impedance source (qZS) inverter seen in Fig. 1.18B is an emerging inverter topology that provides features of CSI and VSI topologies in a single inverter due to the resonance network at the input of the inverter. The operation modes of this inverter determine source configuration of the inverter based on series-connected inductance or parallel-connected capacitance to the supply source. An essential application of multiresonant inverter topologies is illustrated in Fig. 1.18C, where an LLCC series-parallel resonant inverter configuration is presented. This resonant network enhances short-circuit immunity of the inverter while increasing the operating voltage range [7]. The peak efficiency of this inverter topology with an LLCC resonant network can reach 97.4%, as noted in [21]. Another application of a resonant network configured with an LLC topology is presented in Fig. 1.18D, where a dual-mode resonant topology is composed of a conventional H-bridge.

The switching losses occurring in the primary of the HF transformer are eliminated by the resonant circuit and parallel switches on the secondary side. The switching frequency of this circuit has been extended up to 1 MHz with 97.6% overall efficiency [7, 22]. LLC resonant networks are mostly used in various frequency modulation methods for regulating the fluctuation of supply voltage and loads.

1.4 Fundamentals of control schemes

The control and operation of inverters are usually performed by using PWM techniques improved with additional features for eliminating harmonic content and increasing the efficiency. The most widely used methods have been summarized as SPWM, SHE-PWM, and SVM modulation schemes. Although these are the main methods that have been used for many years, a wide range of novel modulation schemes, including additional control methods, can be found in the literature. The aforementioned schemes are assumed as the fundamental controllers while novel controllers are improved with the aid of fuzzy logic controllers, genetic and evolutionary algorithms, soft-computing methods, and several search algorithms.

A general classification of PWM methods used in control of multilevel inverters is listed in Fig. 1.19. The PWM generation can be reference-based or carrier-based comparisons. The reference-based generation methods can be accomplished by using single or multiple references as in carrier-based generation. Both methods are based

CHAPTER 1 Introduction to multilevel voltage source inverters

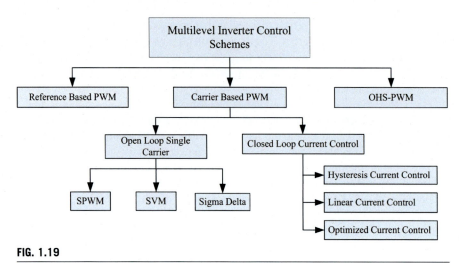

FIG. 1.19
A general list of control schemes used in multilevel inverters.

on comparison procedures followed to generate switching pulses that are mostly obtained as an outcome of an SPWM scheme. One of the widely used SPWM generation methods is based on multiple carrier modulation in which the carrier is either phase shifted or level shifted. The carrier-based SPWM methods are used to generate phase disposition (PD), phase opposition and disposition (POD), and alternative phase opposition and disposition (APOD) based SPWM methods. The carrier signal generation and level- or phase-shifting procedures of these modulation schemes are illustrated according to the aforementioned order in Fig. 1.20A to C, while the phase-shifting carrier based SPWM scheme is shown in Fig. 1.20D [4, 9]. The level-shifting modulation methods are convenient to determine the level of an inverter, where an m-level inverter requires $(m-1)$ number of carrier signals. In this regard, all carriers seen in Fig. 1.20 are generated for a five-level multilevel inverter.

The SHE-PWM method is an alternative to regular SPWM in which the harmonic orders to be eliminated are formulated and switching angles are calculated for certain harmonic components. The modulation method was improved and proposed by Patel in 1964. Each switching angle is determined based on the output level of MLI and predefined harmonic orders by using Fourier series expansion.

In an example 11-level output waveform, the expansion is obtained as seen in Eq. (1.21):

$$V(wt) = \sum_{n=1,3,5,..}^{\infty} \frac{4V_{DC}}{n.\pi}.(\cos(n.\theta_1) + \cos(n.\theta_2) + ... \cos(n.\theta_5)).\sin(n.wt) \quad (1.21)$$

where n denotes the harmonic orders to be eliminated. If the selected harmonics are defined as 5th, 7th, 11th, and 13th orders, the required exact switching angles are determined as follows:

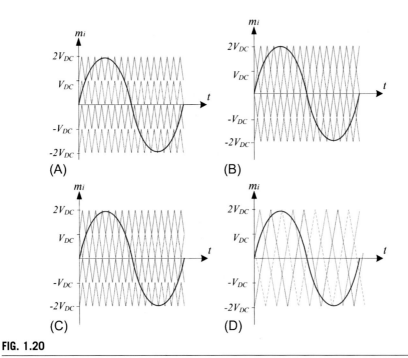

FIG. 1.20

Multicarrier SPWM control strategies: (A) PD, (B) POD, (C) APOD, (D) PS.

$$\cos(\theta_1) + \cos(\theta_2) + \cos(\theta_3) + \cos(\theta_4) + \cos(\theta_5) = 5 \cdot m_a \quad (1.22)$$

$$\cos(5 \cdot \theta_1) + \cos(5 \cdot \theta_2) + \ldots + \cos(5 \cdot \theta_5) = (0)$$

$$\cos(7 \cdot \theta_1) + \cos(7 \cdot \theta_2) + \ldots + \cos(7 \cdot \theta_5) = 0$$

$$\cos(11 \cdot \theta_1) + \cos(11 \cdot \theta_2) + \ldots + \cos(11 \cdot \theta_5) = 0$$

$$\cos(13 \cdot \theta_1) + \cos(13 \cdot \theta_2) + \ldots + \cos(13 \cdot \theta_5) = 0$$

The $\theta_1, \theta_2, \ldots, \theta_5$ coefficients are switching angles for eliminating the targeted harmonic orders and decrease the overall THD ratios. Since the represented values are not linear in these equations, the solution is obtained by using Newton-Raphson iterations. If the modulation index is selected as 0.8, the switching angles are calculated as the values $\theta_1 = 6.57°$, $\theta_2 = 18.94°$, $\theta_3 = 27.18°$, $\theta_4 = 45.14°$, and $\theta_5 = 62.24°$.

The predefined switching angles are written to a look-up table in the application stage and switching pulses are generated by a microprocessor based on these values. The most important drawback of SHE-PWM is the requirement of the switching angle calculation as in the fundamental frequency switching. Although the

Newton-Raphson iteration can calculate the switching angles given in Eq. (1.22), the initial values are obtained by assumptions or predictions that may cause unstable operation at the transients. On the other hand, increased voltage level and source number complicate solving the equation [4, 9].

Another common modulation scheme used in multilevel inverter control is SVM, which is based on reference frame transformation for calculating and generating the switching signals. In SVM, the signal generation is based on reference acquisition and transforming the three-phase rotating frame to a two-phase stationary d-q coordinate. Then, the reference stationary frame is used to generate the α-β complex space. It is noted that the harmonic elimination of the SVM method is much better than SPWM due to the increased value of the triangular carrier. However, SVM is not suitable to be used in higher output levels such as five level or more due to the increased calculation requirements [4]. The modulation and control methods are presented as a summary in this chapter since they are analyzed in more detail in the following chapters.

References

[1] Q.-C. Zhong, T. Hornik, Control of Power Inverters in Renewable Energy and Smart Grid Integration: Zhong/Control, John Wiley & Sons, Ltd., Chichester, West Sussex, United Kingdom, 2012. https://doi.org/10.1002/9781118481806.

[2] F.L. Luo, H. Ye, Power Electronics Advanced Conversion Technologies, CRC Press, 2018.

[3] M. Rashid, Power Electronics Handbook, Academic Press, 2001.

[4] I. Colak, E. Kabalci, R. Bayindir, Review of multilevel voltage source inverter topologies and control schemes, Energ. Conver. Manage. 52 (2011) 1114–1128. https://doi.org/10.1016/j.enconman.2010.09.006.

[5] B.K. Bose, Modern Power Electronics and Ac Drives.pdf, Prentice Hall PTR, 2002.

[6] A. Nabae, I. Takahashi, H. Akagi, A new neutral-point-clamped PWM inverter, IEEE Trans. Ind. Appl. IA-17 (1981) 518–523. https://doi.org/10.1109/TIA.1981.4503992.

[7] E. Kabalcı, Review on novel single-phase grid-connected solar inverters: circuits and control methods, Sol. Energy. 198 (2020) 247–274. https://doi.org/10.1016/j.solener.2020.01.063.

[8] A. Sinha, K. Chandra Jana, M. Kumar Das, An inclusive review on different multi-level inverter topologies, their modulation and control strategies for a grid connected photovoltaic system, Sol. Energy. 170 (2018) 633–657. https://doi.org/10.1016/j.solener.2018.06.001.

[9] N. Prabaharan, K. Palanisamy, A comprehensive review on reduced switch multilevel inverter topologies, modulation techniques and applications, Renew. Sustain. Energy Rev. 76 (2017) 1248–1282. https://doi.org/10.1016/j.rser.2017.03.121.

[10] J. Rodriguez, L.G. Franquelo, S. Kouro, J.I. Leon, R.C. Portillo, M.A.M. Prats, M.A. Perez, Multilevel converters: an enabling Technology for High-Power Applications, Proc. IEEE 97 (2009) 1786–1817. https://doi.org/10.1109/JPROC.2009.2030235.

[11] A. Shukla, A. Ghosh, A. Josh, Flying capacitor multilevel inverter and its applications in series compensation of transmission lines, in: IEEE Power Eng. Soc. Gen. Meet, vol. 2004, IEEE, Denver, CO,USA, 2004, pp. 1453–1458. https://doi.org/10.1109/PES.2004.1373109.

[12] J.-S. Lai, F. Zheng, Peng, multilevel converters-a new breed of power converters, IEEE Trans. Ind. Appl. 32 (1996) 509–517. https://doi.org/10.1109/28.502161.
[13] F.Z. Peng, J.W. McKeever, J. VanCoevering, A multilevel voltage-source inverter with separate DC sources for static VAr generation, IEEE Trans. Ind. Appl. 32 (1996) 9.
[14] I. Colak, E. Kabalci, R. Bayindir, S. Sagiroglu, The Design and Analysis of a 5-Level Cascaded Voltage Source Inverter with Low THD, in: 2009 Int, Conf. Power Eng. Energy Electr. Drives, IEEE, Lisbon, Portugal, 2009, pp. 575–580. https://doi.org/10.1109/POWERENG.2009.4915185.
[15] E. Kabalci, I. Colak, R. Bayindir, C. Pavlitov, in: Modelling a 7-Level Asymmetrical H-Bridge Multilevel Inverter with PS-SPWM Control, Int. Aegean Conf. Electr. Mach. Power Electron. Electromotion Jt. Conf, IEEE, Istanbul, Turkey, 2011, pp. 578–583. https://doi.org/10.1109/ACEMP.2011.6490663.
[16] I. Colak, E. Kabalci, G. Keven, Simulation of a seven-level asymmetric Cascade multilevel inverter with PR control, in: 16th Int. Power Electron. Motion Control Conf. Expo., Antalya Turkey, 2014, p. 5.
[17] E. Kabalci, Y. Kabalci, R. Canbaz, G. Gokkus, Single Phase Multilevel String Inverter for Solar Applications, Palermo, Italy, 2015, pp. 109–114.
[18] I. Colak, E. Kabalci, R. Bayindir, G. Bal, in: Modeling of a Three Phase SPWM Multilevel VSI with Low THD Using Matlab/Simulink, IEEE, 2009 13th European Conference on Power Electronics and Applications, 2009.
[19] I. Colak, M. Demirtas, E. Kabalci, Design, optimisation and application of a resonant DC link inverter for solar energy systems, COMPEL - Int. J. Comput. Math. Electr. Electron. Eng. 33 (2014) 1761–1776. https://doi.org/10.1108/COMPEL-06-2013-0200.
[20] M.K. Kazimierczuk, Resonant power converters, Wiley IEEE, 2011.
[21] D. Vinnikov, A. Chub, E. Liivik, I. Roasto, High-performance quasi-Z-source series resonant DC–DC converter for photovoltaic module-level power electronics applications, IEEE Trans. Power Electron. 32 (2017) 3634–3650. https://doi.org/10.1109/TPEL.2016.2591726.
[22] R. Hasan, S. Mekhilef, M. Seyedmahmoudian, B. Horan, Grid-connected isolated PV microinverters: a review, Renew. Sustain. Energy Rev. 67 (2017) 1065–1080. https://doi.org/10.1016/j.rser.2016.09.082.

CHAPTER 2

Neutral-point-clamped and T-type multilevel inverters

Hasan Komurcugil[a] and Sertac Bayhan[b]

[a]*Department of Computer Engineering, Eastern Mediterranean University, Famagusta, Mersin 10, Turkey* [b]*Qatar Environment and Energy Research Institute, Hamad Bin Khalifa University, Doha, Qatar*

2.1 Neutral-point-clamped multilevel inverters

Although semiconductor switches are enough for generating two-level voltage at the output of the traditional voltage source inverters (VSIs), the multilevel inverters need some additional components—either passive or active—to generate multilevel output voltages. The diode-clamped multilevel inverters use diodes and capacitors for this purpose. The output voltage level is determined by the count and configuration of active switches, diodes, and capacitors. Although various configurations are possible to obtain multilevel output voltage, the three-level inverter, often known as the neutral-point-clamped (NPC) inverter, has found wide application [1, 2].

In addition to the traditional NPC inverter topology, many derived forms of NPC inverters have been developed. For example, the active-neutral-point-clamped (ANPC) inverter, which is an arrangement of two-level inverters connected in series, is proposed in Ref. [3]. The proposed ANPC inverter is based on the combination of NPC and floating capacitor converters. In Ref. [4], a single-phase cascaded H-bridge NPC inverter is used as a single-phase grid-connected inverter. The topology comprises two H-bridge NPC inverters in series that result in a five-level output voltage. The main advantage of this topology is low total harmonic distortion (THD) and low filter size requirement. The NPC inverter is also used in the single-stage photovoltaic (PV) inverters [5, 6]. The advantages of the NPC/H-bridge inverter and impedance source inverters (qZSI) are combined to obtain a unique power converter topology for high-power PV applications in Refs. [5–7]. A new family of seven-level boost active-neutral-point-clamped (7 L-BANPC) circuits is proposed in Ref. [8]. The proposed topology is derived from the conventional ANPC by adding one additional capacitor to boost the voltage and increase the number of output voltage levels.

Similar to other multilevel inverters, the main advantages of the NPC inverters over the traditional VSIs are low *dv/dt* and THD values. In this section, the NPC inverters are investigated. First, the converter configuration and switching states are presented. After that, modulation techniques were summarized. Finally, a model predictive current control technique for the NPC inverter is explained in the last part of this section.

2.1.1 Converter configuration

Fig. 2.1 illustrates the circuit configuration of a three-level NPC inverter [2]. Each inverter leg is composed of four switching devices (S_1–S_4)—either IGBT or MOSFET—along with four antiparallel body diodes. The dc-link capacitor is split into two parts that provide a neutral point on the dc side of the inverter. The diodes connected to the neutral point, D_1 and D_2, are the clamping diodes. The output voltage "a" can be of three values: $+E$, 0, and $-E$. The switches S_{1a} and S_{2a} are turned on for voltage $+E$; the switches S_{3a} and S_{4a} are turned on for voltage $-E$; and the switches S_{2a} and S_{3a} are turned on for $a = 0$. The diodes D_{1a} and D_{2a} clamp the switch voltages to E or $-E$ when S_{1a} and S_{2a} are turned on or when S_{3a} and S_{4a} are turned on.

The voltage stress on each switching device in this configuration is only one capacitor voltage ($E = V_s/N$), where V_s is the dc-link voltage. This converter topology, theoretically, can be extended to any voltage level in high-voltage applications by using low-voltage devices. On the other hand, each phase needs (N–1) x (N–2) additional clamping diodes to provide an N-level voltage. A large number of clamping diodes result in high cost and packaging problems for high-voltage level applications. Furthermore, to balance the capacitor voltages in high-level NPC inverter applications, a special control technique is mandatory, which will be discussed in the later sections. Thus, the majority of practical applications for a diode-clamped multilevel inverter are limited to below five levels [2, 9].

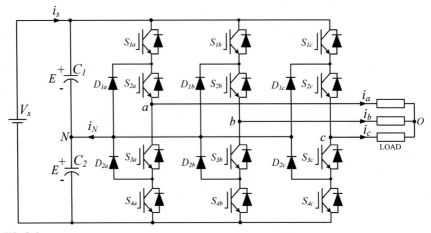

FIG. 2.1

Three-level neutral-point-clamped inverter configuration.

2.1.2 Switching states and commutation

The switching states of a single-phase NPC inverter are given in Table 2.1. The inverter terminal voltage (v_{aN}) is $+E$ when the upper two switches (S_1 and S_2) are turned on (switching state "P") while the v_{AN} is $-E$ when the lower two switches (S_3 and S_4) are turned on (switching state "N"). Furthermore, the inverter terminal voltage (v_{aN}) is 0 when the inner switches (S_2 and S_3) are turned on (switching state "O"). It can be observed from Table 2.1 that switches S_1 and S_3 operate in a complementary manner [1]. It means that if S_1 is switched on, S_3 must be switched off, and vice versa. Similarly, S_2 and S_4 are a complementary pair as well.

The switching states, gate signals, and inverter terminal voltage are depicted in Fig. 2.2. The gate signals (v_{g1}–v_{g4}) are applied to the switches (S_{1a}–S_{4a}) in Fig. 2.1. To generate gate signals, various modulation techniques similar to traditional VSI—such as sinusoidal pulse width modulation (SPWM), space vector PWM, selective harmonic elimination schemes, etc., can be used. As shown in Fig. 2.2, the inverter terminal voltage has three voltage levels "$+E$," "0," and "$-E$," based on which the inverter is referred to as a three-level inverter [1].

Table 2.1 Switching states of a single-phase NPC inverter.

Switching state	Device switching status (Phase a)				Inverter terminal voltage (v_{aN})
	S_1	S_2	S_3	S_4	
P	ON	ON	OFF	OFF	$+E$
O	OFF	ON	ON	OFF	0
N	OFF	OFF	ON	ON	$-E$

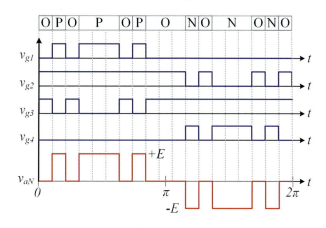

FIG. 2.2

Switching states, gate signals, and inverter terminal voltage (v_{aN}).

Now, let us consider a transition from the switching state "O" to "P," so as to examine the commutation of switching devices in the NPC inverter. To do that, we need to turn S_1 on and turn S_3 off, as shown in Fig. 2.3A. It is clear that S_1 and S_3 and S_2 and S_4 are complementary switch pairs. For that reason, a dead time (δ) is required between transition switch pairs similar to the traditional VSIs.

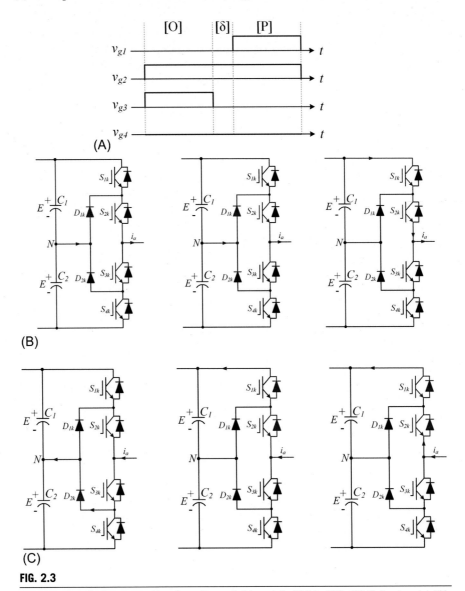

FIG. 2.3

Commutation during a transition from the switching state "O" to "P". (A) Gate signals, (B) commutation with $i_a > 0$, and (C) commutation with $i_a < 0$.

The phase current (i_a) paths under different commutation conditions on the first leg are shown in Fig. 2.3B and C [1]. Two cases are considered as follows;

Case 1: In this case, we consider the load current "i_a" is positive and constant due to the inductive characteristic load. We also assumed that the dc-link capacitors (C_1 and C_2) are quite large so the voltage amplitude on each capacitor is "E" and all the components in the NPC inverter configuration are ideal. The commutation steps are shown in Fig. 2.3B and the active devices are highlighted in blue color. It can be seen that, during the switching state "O," S_{1k} and S_{4k} are switched off, whereas S_{2k} and S_{3k} are switched on. Furthermore, the clamping diode "D_{1k}" is in conduction mode in this interval due to the positive load current ($i_a > 0$). It is clear that the voltage drops on S_{2k} and S_{3k} are zero and S_{1k} and S_{4k} are equal to the capacitor voltage "E." As mentioned before, we need a dead time during transition from the switching state "O" to the switching state "P" and vice versa. During this period, S_{3k} is turned off while S_{2k} is kept on. The current path will be exactly the same as the previous current path. On the other hand, the voltage drops on the S_{3k} and S_{4k} become half of the capacitor voltage "$E/2$." The top switch S_{1k} is turned on while S_{2k} is kept on during the switching state "P." In this switching state, the clamping diode "D_{1k}" is in cutoff mode since it is reverse-biased. It can be seen that the voltages on S_{1k} and S_{2k} are zero, whereas the voltages on the S_{3k} and S_{4k} are equal to the capacitor voltage "E." Please note that the current path remains the same [1].

Case 2: In this case, the load current "i_a" is considered as negative and constant due to the inductive characteristic load. The commutation steps are shown in Fig. 2.3C and the active devices are highlighted in blue color. In the switching state "O," S_{2k} and S_{3k} are turned on and the clamping diode "D_{2k}" is turned on by the negative load current "i_a." The voltage drops on the S_{1k} and S_{4k} are equal to the capacitor voltage "E." During the "δ" internal, S_{3k} is turned off while S_{2k} is kept on similar to the previous case. During this interval, the voltage drops on S_{1k} and S_{2k} are zero and S_{3k} and S_{4k} are equal to the capacitor voltage "E." It is clear that the load current "i_a" is commutated from S_{3k} to the diodes. Although the top switches S_{1k} and S_{2k} are in the conduction mode in the switching state "P," these switches do not carry the load current because of the conduction of antiparallel body diodes. During this switching state, the voltage drops on S_{1k} and S_{2k} are zero and S_{3k} and S_{4k} are equal to the capacitor voltage "E."

It can be concluded that all the switching devices in the NPC inverter withstand only half of the dc bus voltage during the commutation from the switching state "O" to "P." Similarly, the same conclusion can be drawn for the commutation from [P] to [O], [N] to [O], or vice versa. Therefore, the switches in the NPC inverter do not have a dynamic voltage sharing problem. It should be pointed out that the switching between [P] and [N] is prohibited for two reasons: (a) It involves all four switches in an inverter leg, two being turned on and the other two being commutated off, during which the dynamic voltage on each switch may not be kept the same; and (b) the switching loss is doubled [1].

2.1.3 Modulation techniques

The modulation techniques, in general, for NPC converters can be divided into two mainframes:

2.1.3.1 Sinusoidal pulse width modulation (SPWM)

The SPWM mainly is employed in industrial applications and based on the comparison of modulation and carrier signals. A sine wave (modulation signal, v_m) is compared with two triangular waveforms (carrier signals, v_{c1} and v_{c2}) to generate PWM signals as shown in Fig. 2.4. It should be mentioned that this modulation scheme is only represented for the phase "a." To generate switching signals for the other phases, modulation signals should be shifted 120° according to each other while using the same carrier signals. The frequency of the modulation signal determines the output voltage frequency while the frequency of the carrier signals determines the switching frequency. Furthermore, the amplitude of the output voltage is determined by the amplitude of the modulation signal.

Similar to the traditional two-level SPWM, the amplitude of the first harmonic of the voltage supplied by the inverter is proportional to the amplitude of the modulating signal only if this latter does not exceed amplitude V_c of the carrier. This

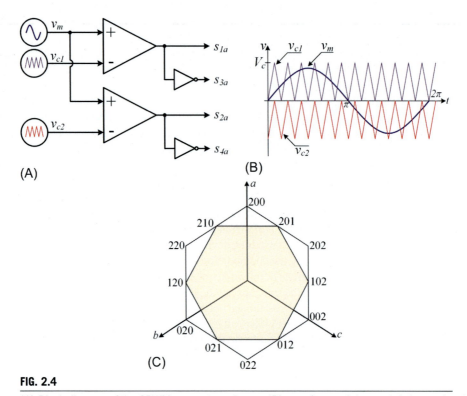

FIG. 2.4

(A) Block diagram of the SPWM generator scheme; (B) waveforms of the modulation and carrier signals; and (C) hexagon containing the possible voltage vectors for SPWM.

limitation implies that using SPWM, a voltage vector can be realized only if it is inside the inner hexagon depicted in Fig. 2.4C. Therefore, in the steady-state sinusoidal operation, the voltage representative vector can assume a maximum value equal to 0.75 V_{max} [10].

2.1.3.2 Space vector pulse width modulation (SV-PWM)

Although the space vector PWM modulation (SV-PWM) technique was initially proposed as a vector approach for three-phase traditional inverters, several SVM techniques have been derived from its original form and utilized in a number of multilevel inverter topologies [9] The main advantage of SV-PWM over SPWM is that this technique provides a higher voltage to the load with lower total harmonic distortion (THD).

In Ref. [10], various SV-PWM techniques were proposed for NPC inverters. The technique in Ref. [4] considered all advantages of the traditional SV-PWM that include ensuring commutations only between adjacent states. Furthermore, this technique guarantees that only three commutations are affected at each modulation interval and all the voltage vectors in the hexagon can be realized as shown in Fig. 2.5A. To ensure the advantages of SV-PWM for NPC inverters, the output voltage plane, as shown in Fig. 2.5A, is divided into six overlapped hexagons as depicted in Fig. 2.5B. The highlighted part in Fig. 2.5B represents two hexagons' coinciding parts. The objectives of dividing the plane into subdivisions can be listed as follows: (1) the application, inside each hexagon, of the standard space vector modulation used in two-level inverters; (2) make as easy as possible the transition from a hexagon to the subsequent one [10].

To implement SV-PWM in the digital platforms (microcontrollers, FPGAs), certain steps need to be followed. First of all, the voltage space vector needs to be identified by two integer variables. Here we selected "i" and "n" as the integer variables. The variable "i" states the actual hexagon "h_{ei}" as shown in Fig. 2.5B, whereas the variable "n" specifies the voltage vector "v_n" inside the hexagon as shown in Fig. 2.5C and D.

The location of the voltage vector depends on the operating conditions, and two main methods can be used to identify the order of the voltage vectors; (1) the modulation stays inside the same hexagon as shown in Fig. 2.5C in case the desired voltage vector does not overcome a fixed angular threshold; (2) on the other hand, a suitable procedure must be applied to move the modulation to an adjacent hexagon as shown in Fig. 2.5D.

2.1.4 Finite set model predictive control of a three-phase three-level neutral-point-clamped inverter

Although a number of control techniques have been presented for the NPC inverter, in this section, a model predictive control (MPC)-based control technique is presented. Recently, the MPC technique has become a mature control technique for

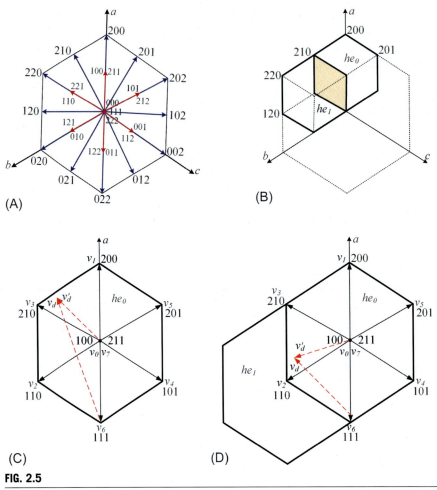

FIG. 2.5

(A) NPC inverter output voltage vectors; (B) subdivision of the voltage plane into hexagons; (C) modulation inside hexagon h_{e0}; and (D) transition from hexagon h_{e0} to hexagon h_{e1}.

power electronics similar to PI-based control techniques. MPC solves the optimization problem at each discrete time to find the optimal inverter switching state. This optimization problem refers to the minimization of the cost function that represents the expected behavior of the system over a finite horizon. The major advantages of MPC are fast dynamic response, intuitive handling of multivariables, nonlinearities, and easy inclusion of system constraints [11]. In general, MPC takes the constraints directly into the cost function that is imposed to simplify the optimization problem. MPC evaluates all possible switching combinations to obtain the optimal solution that minimizes the cost function [12].

2.1.4.1 System model

Three-phase three-level NPC inverter topology with RL load is given in Fig. 2.6 Each leg of the inverter consists of four power switches with two diodes that allow the output terminal to be connected to the middle point of the DC-link capacitors. This circuit configuration allows to generate three voltage levels at the output of the inverter according to the neutral point. The switching combinations are given in Table 2.2 [13]. Switching state variable S_x represents the switching state of phase x, with $x = a; b; c$, and it has three possible values denoted by positive, zero, and negative that represent the switching combinations that generate $V_{dc}/2$, 0, and $-V_{dc}/2$, respectively, at the output of the inverter phase.

Fig. 2.7 illustrates possible voltage vectors and switching states generated by a three-level NPC inverter. It can be seen that there are 19 voltage vectors that are generated from 27 switching states [14]. The continuous-time expression for the output current ($\mathbf{i_o}$) can be expressed as

$$L\frac{d\mathbf{i}(t)}{dt} = \mathbf{v}(t) - R\mathbf{i}(t) - \mathbf{e}(t) \quad (2.1)$$

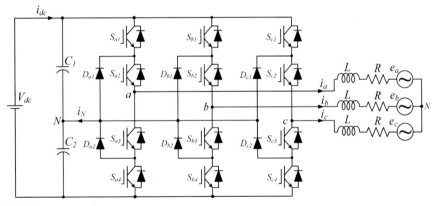

FIG. 2.6

Three-phase three-level NPC inverter topology with RL load.

Table 2.2 Switching states for one phase of the NPC inverter.

S_x	S_{x1}	S_{x2}	S_{x3}	S_{x4}	v_{xN}
+	1	1	0	0	$V_{dc}/2$
0	0	1	1	0	0
−	0	0	1	1	$-V_{dc}/2$

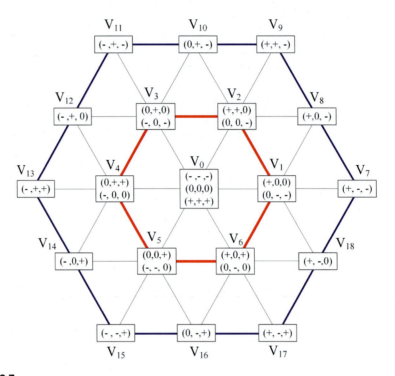

FIG. 2.7

Possible voltage vectors and switching states generated by a three-level NPC inverter.

where R and L are the load resistance and inductance, respectively, $\mathbf{v}$ is the voltage vector generated by the NPC inverter, $\mathbf{e}$ is the electromotive force of the load, and $\mathbf{i}$ is the load current vector. These vectors are defined as

$$\left.\begin{aligned} \mathbf{v} &= \frac{2}{3}\left(V_a + \mathbf{a}V_b + \mathbf{a}^2 V_c\right) \\ \mathbf{i} &= \frac{2}{3}\left(i_a + \mathbf{a}i_b + \mathbf{a}^2 i_c\right) \\ \mathbf{e} &= \frac{2}{3}\left(e_a + \mathbf{a}e_b + \mathbf{a}^2 e_c\right) \end{aligned}\right\} \quad (2.2)$$

where $a = e^{j(2\pi/3)}$.

2.1.4.2 Model predictive current control

MPC uses a system model to predict the future behavior of the controlled variables over a prediction horizon. The predicted values are used by the controller to obtain the optimal actuation, according to a predefined cost function. Predictive control gives the possibility to avoid a cascaded structure, which is typically used in the linear control scheme. The simplified block diagram of the MPC technique for the three-phase NPC inverter is depicted in Fig. 2.8. The MPC technique mainly contains of two main blocks: (1) a predictive model and (2) minimization of the cost function.

A. A Predictive Model of the NPC

This section describes how to obtain predictive models from continuous-time equations. To control the output current vector (**i**) of the inverter, the predictive model must be expressed into the discrete-time model. One way to obtain a discrete-time model is to use the forward-difference Euler equation due to its simplicity [13]. The derivative form $dx(t)/dt$ is approximated by

$$\frac{dx(t)}{dt} \approx \frac{x(k+1) - x(k)}{T_s} \quad (2.3)$$

where T_s is the sampling time, x is the control parameter, in this case, the output current vector (**i**). By substituting Eqs. (3.2) into (3.3), the discrete-time model of the output current vector is

$$\mathbf{i}(k+1) = \frac{T_s}{RT_s + L}\left[\frac{L}{T_s}\mathbf{i} + \mathbf{v}(k+1) - \mathbf{e}(k+1)\right] \quad (2.4)$$

where **i** $(k+1)$ is the predicted output current vector at the next sampling time. The current prediction in Eq. (3.4) also requires an estimation of the future load back EMF **e** $(k+1)$. It can be estimated by using a second-order extrapolation from present and past values. By using Eq. (3.3), the output voltage and current can be expressed as:

$$\hat{\mathbf{e}}(k) = \mathbf{v}(k) + \frac{L}{T_s}\mathbf{i}(k-1) - \frac{RT_s + L}{T_s}\mathbf{i}(k) \quad (2.5)$$

The same approximation can be considered for capacitor voltages [14]. The discrete-time equations for capacitor voltages are:

$$\left.\begin{array}{l}v_{C1}(k+1) = v_{C1}(k) + \frac{1}{C}i_{C1}(k)T_s \\ v_{C2}(k+1) = v_{C2}(k) + \frac{1}{C}i_{C2}(k)T_s\end{array}\right\} \quad (2.6)$$

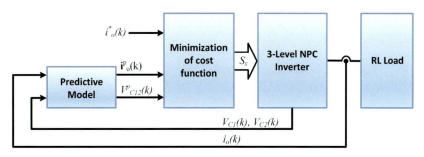

FIG. 2.8

The simplified block diagram of the model predictive control technique.

where currents $i_{c1}(k)$ and $i_{c2}(k)$ depend on the inverter switching state. The capacitor currents can be calculated using the following expressions:

$$\left.\begin{array}{l}i_{C1}(k+1)=i_{dc}-H_{1a}i_a(k)-H_{1b}i_b(k)-H_{1c}i_c(k)\\ i_{C2}(k+1)=i_{dc}-H_{2a}i_a(k)-H_{2b}i_b(k)-H_{2c}i_c(k)\end{array}\right\} \quad (2.7)$$

where i_{dc} is the dc source current and H_{1x} and H_{2x} depend on the switching states and can be expressed as

$$\left.\begin{array}{l}H_{1x}=\begin{cases}1 \text{ if } S_x=+\\ 0 \text{ otherwise}\end{cases}\\ H_{1x}=\begin{cases}1 \text{ if } S_x=-\\ 0 \text{ otherwise}\end{cases}\end{array}\right\} \quad (2.8)$$

where x is phase a, b, and, c.

B. Cost Function Optimization.

The cost function optimization is a critical part of the MPC technique. All control requirements need to be integrated into the cost function. In this case, the control requirements of the three-phase NPC inverter are the output current and dc-link voltage control. These requirements should be formulated in the form of the cost function to be minimized. The cost function of the NPC inverter can be expressed in:

$$g = \overbrace{\left(\left|i_\alpha^* - i_\alpha^p\right| + \left|i_\beta^* - i_\beta^p\right|\right)}^{\text{load current control}} + \overbrace{\left(\lambda\left|v_{C1}^p - v_{C2}^p\right|\right)}^{\text{dc–link voltage control}} \quad (2.9)$$

The objective of the first term is to minimize the error between the reference currents (i_α^* and i_β^*) and the predicted output currents (i_α^p and i_β^p) in the orthogonal coordinate axis. The second term in the cost function measures the difference in the predicted values of the DC link capacitor voltages.

2.1.4.3 Simulation results

To validate the performance of the MPC-based control technique, a model of the system depicted in Fig. 2.6 is constructed in the Matlab/Simulink environment. The parameters of the system are summarized in Table 2.3.

The simulation results of the steady-state operation are shown in Fig. 2.9. The reference output current is set to 20 A and it is clear that the output current tracks

Table 2.3 Simulation parameters.

Parameter	Value
Input DC voltage (V_{dc})	1000 V
Capacitors (C_1 and C_2)	470 µF
Load inductance (L)	2 mH
Load resistance (R)	10 Ω
Nominal Frequency	50 Hz
Sampling Time (T_s)	50 µs

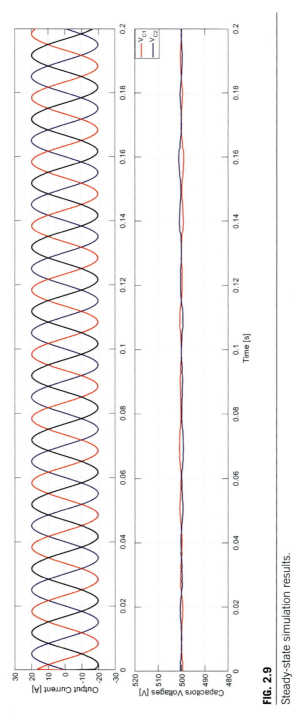

FIG. 2.9

Steady-state simulation results.

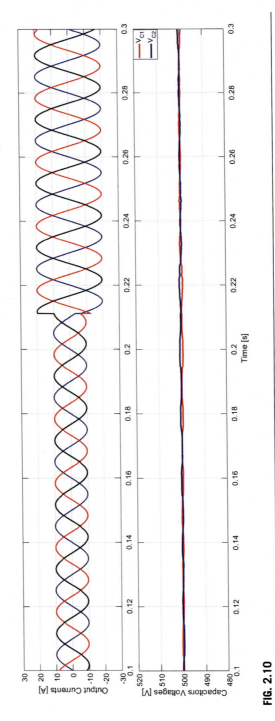

FIG. 2.10

Dynamic response simulation studies.

its reference with high accuracy while the dc-link capacitor voltages are balanced at 500 V.

The dynamic response results are given in Fig. 2.10. The reference current is a step up from 10 to 20 A as shown in Fig. 2.10. It can be observed from the results that the transient time is very short and there is no overshoot. Furthermore, the capacitor voltages are balanced during the dynamic response.

2.2 T-type inverter

T-type inverters are the new generation of multilevel inverters offering better efficiency than the NPC inverters when they are operated within the medium switching frequency range (6–30 kHz) [15, 16] . The improvement in the efficiency is due to the small conduction and switching losses of the T-type topology. Another advantage of the T-type inverters is that, unlike the NPC inverters, no clamping diodes are required for clamping the neutral point to positive or negative dc voltages. In T-type inverters, clamping is achieved by using an active bidirectional switching device connected between the midpoint of each phase leg and midpoint of series connected dc-link capacitors. The bidirectional switching device consists of two anti-series insulated-gate bipolar transistors (IGBTs) which are connected in common-emitter (CE) or in common-collector (CC) configurations as shown in Fig. 2.11. While the CE configuration requires an additional isolated gate drive supply voltage for each phase leg (i.e., three voltage supplies for a three-phase inverter), the CC configuration requires only one isolated supply since the emitters of S_{3k} are common and S_{2k} shares a common emitter with S_{1k} [17, 18]. However, like NPC inverters, T-type inverters also suffer from voltage imbalance existing in dc-link capacitors. The operating principle, switch fault, influence of switching states on the dc capacitor voltages and modulation of the T-type inverter topology are discussed in the following subsections.

2.2.1 Description of T-type inverter and its operating principle

The topologies of a three-phase three-level T-type inverter with CE and CC configurations are depicted in Fig. 2.11. There are four switching devices per leg. It is clear from Fig. 2.11 that the three-level T-type inverter topology is an extended version of the conventional two-level inverter with an additional active bidirectional switch connected between each phase leg and the dc-link midpoint. While switches S_{1k} and S_{4k} ($k = a, b, c$) block the full dc-link voltage (V_s), switches S_{2k} and S_{3k} block only half of the dc-link voltage ($V_s/2$). This means that the switching devices having a lower voltage rating can be used in a three-level T-type inverter unlike the NPC inverter topology [19]. Also, it is obvious from Fig. 2.11 that the three-level T-type inverter topology does not employ clamping diodes, which means that it has a smaller part count compared to the three-level NPC inverter topology. Hence, compared to the three-level NPC inverter, the power losses in a three-level T-type inverter are

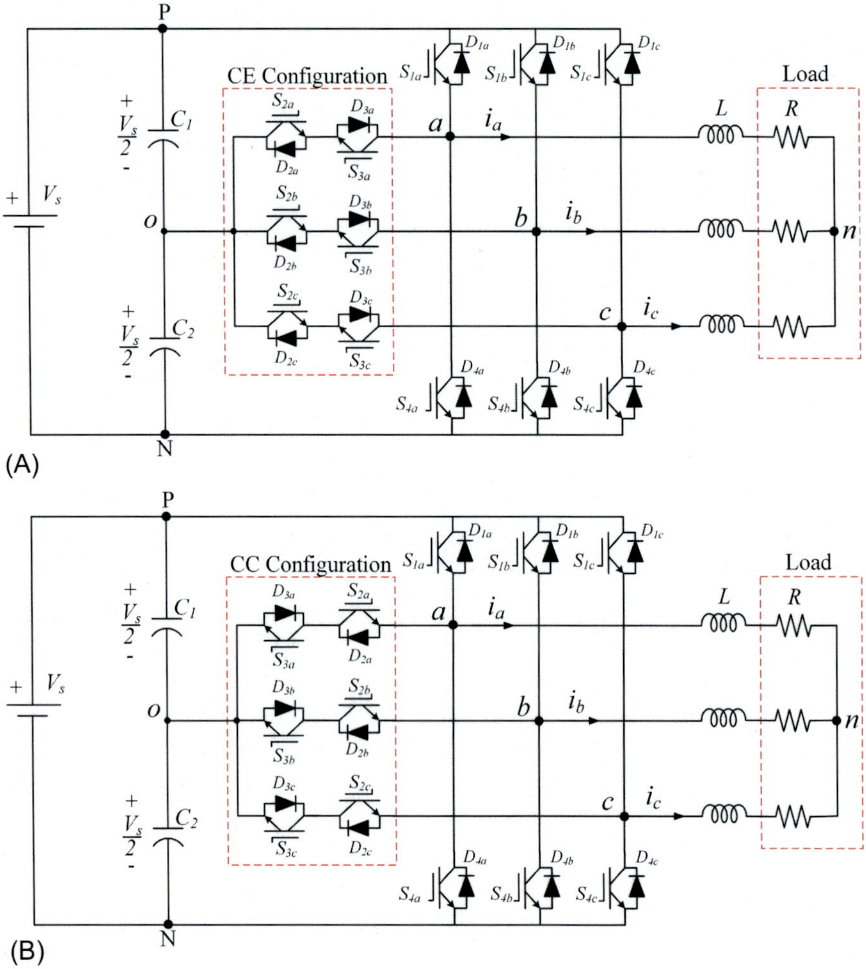

CHAPTER 2 Neutral-point-clamped and T-type multilevel inverters

FIG. 2.11

Three-phase three-level T-type inverter with CE and CC configurations: (A) CE configuration and (B) CC configuration.

reduced considerably at lower switching frequencies [20]. Considering the facts mentioned above, the T-type inverter offers a lower total harmonic distortion (THD) than that of the NPC inverter [15].

There are 12 switches in the three-phase T-type inverter as shown in Fig. 2.11. Using different combinations of these switches, it is possible to connect the output of each phase leg (midpoint of the phase leg) to the positive (P), neutral (0), or negative (N) dc-link voltage levels. Table 2.4 presents the operating states, switching states, and generated output voltages with respect to the neutral point 0. It should be noted

2.2 T-type inverter

Table 2.4 Operating states, switching states and pole voltages.

Operating State	S_{1k}	S_{2k}	S_{3k}	S_{4k}	v_{k0}
P	ON	ON	OFF	OFF	$+V_s/2$
0	OFF	ON	ON	OFF	0
N	OFF	OFF	ON	ON	$-V_s/2$

that S_{3k} is always a complement of S_{1k} and S_{4k} is always a complement of S_{2k}. Considering the switch states in the first row of Table 2.4 (S_{1k} and S_{2k} are ON and S_{3k} and S_{4k} are OFF) and Fig. 2.11, the inverter operates in the P state and generates a pole voltage equal to $v_{k0} = +V_s/2$. Similarly, the switch states in the second row of Table 2.4 (S_{2k} and S_{3k} are ON and S_{1k} and S_{4k} are OFF) cause the inverter to operate in the 0 state generating 0 V at the output. Finally, when the switch states are as in the third row of Table 2.4 (S_{1k} and S_{2k} are OFF and S_{3k} and S_{4k} are ON), the inverter operates in the N state and generates an output voltage equal to $v_{k0} = -V_s/2$.

Since the output current i_k is either positive or negative, each phase leg can be described by six different paths depending on the operating states (P, 0, and N) [15]. When the output current is positive ($i_k > 0$), the inverter operates in one of the operating modes presented in Table 2.4 and the possible paths for the output current are as shown in Fig. 2.12A–C. On the other hand, when the output current is negative ($i_k < 0$), the possible paths of the output current at each operating state are as shown in Fig. 2.13A–C. When the output current is positive, it flows through S_{1k} in the P state (see Fig. 2.12A), through S_{2k} and D_{3k} in the 0 state (see Fig. 2.12B), and through D_{4k} in the N state (see Fig. 2.12C). On the other hand, when the output current is negative, it flows through D_{1k} in the P state (see Fig. 2.13A), through S_{3k} and D_{2k} in the 0 state (see Fig. 2.13B), and S_{4k} in the N state (see Fig. 2.13C).

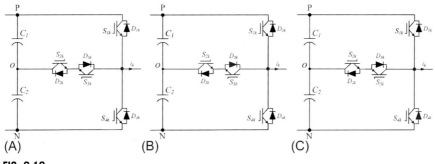

FIG. 2.12

Output current paths for $i_k > 0$ in operating states: (A) P state, (B) 0 state, and (C) N state.

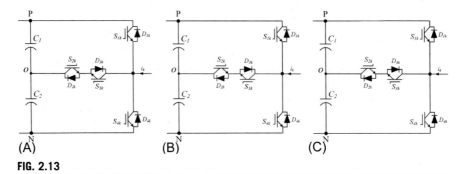

FIG. 2.13

Output current paths for $i_k < 0$ in operating states: (A) P state, (B) O state, and (C) N state.

2.2.2 Switch open-circuit fault

The number of switching devices in a three-level T-type inverter is twice as high as that in a traditional two-level inverter. Hence, the swiching device fault possibility in a three-level T-type inverter is twice as high as that in a two-level inverter. Switching device faults are categorized as switch-open circuit (SOC) fault and switch short-circuit (SSC) fault. In this section, open-circuit fault is discussed. An open-circuit fault occurs due to the high collector current, gate driver fault, and breaking of bonding wires because of thermic cycling [21]. When an open-switch fault occurs, the switching state cannot reach the required state which results in a change in the pole voltage v_{k0} and path of the output current i_k [22,23]. Since an open-switch fault can cause other problems in other parts of the inverter, it should be detected as fast as possible. The output current paths under an open-circuit fault are shown in Fig. 2.14. Comparing the output current paths under normal operation (see Figs. 3.12 and 3.13) and open-circuit fault operation, one can see that the output current path is changed when an open-circuit fault occurs. While a solid line denotes the output current path under normal operation, the dotted line represents the output current path under open-circuit fault. The switching device which has an open-circuit fault is denoted by red color.

Fig. 2.14A shows the output current path for an open-circuit fault at S_{1k} under operating state P when $i_k > 0$. Note that the output current flows through S_{1k} before the open-circuit fault. When an open-circuit fault occurs, the positive dc level does not exist anymore and the output current flows through S_{2k} and D_{3k} instead of S_{1k}. In this case, the midpoint of the vertical inverter leg is connected to the neutral point 0 which changes the pole voltage from $+V_s/2$ to 0. Also, a larger neutral point current flows compared to the normal operation. Hence, V_{C1} becomes larger than V_{C2}. Fig. 2.14B shows the output current path for an open-circuit fault at S_{2k} under operating state 0 when $i_k > 0$. Note that the output current flows through S_{2k} and D_{3k} before the open-circuit fault. When an open-circuit fault occurs, the midpoint of the vertical inverter leg is connected to the negative dc bus, which changes the pole voltage from 0 to $-V_s/2$ and the output current flows through D_{4k} instead of S_{2k} and

2.2 T-type inverter

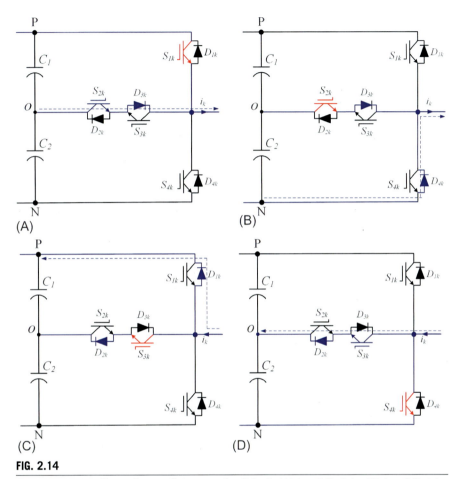

FIG. 2.14

Output current paths under a switch open-circuit fault: (A) $i_k > 0$ P state, (B) $i_k > 0$ 0 state, (C) $i_k < 0$ 0 state, and (D) $i_k < 0$ N state.

D_{3k}. The magnitude of the current flowing through the neutral point is smaller compared to that under normal operation. For this reason, V_{C2} becomes larger than V_{C1}. The output current path for an open-circuit fault at S_{3k} under operating state 0 when $i_k < 0$ is depicted in Fig. 2.14C. The output current flows through S_{3k} and D_{2k} before the open-circuit fault. When an open-circuit fault exists, the midpoint of the vertical inverter leg is connected to the positive dc bus, changing the pole voltage from 0 to $+V_s/2$, and the output current flows through D_{1k} instead of S_{3k} and D_{2k}. In this case, a larger neutral point current flows, which makes V_{C1} larger than V_{C2}. Fig. 2.14D shows the output current path for an open-circuit fault at S_{4k} under operating state N when $i_k < 0$. Note that the output current flows through S_{4k} before the open-circuit fault. After the one-circuit fault takes place, the midpoint of the vertical inverter leg is connected to the neutral point 0, which results in a change in the pole voltage from $-V_s/2$ to 0, and the output current flows through S_{3k} and D_{2k} instead of S_{4k}. Since the neutral point current is smaller, it leads to a larger V_{C2} than V_{C1}.

2.2.3 Switch short-circuit fault

When a switch short-circuit fault occurs, an extremely large current exists which damages the other parts in the system. Therefore, in order to ensure the safety of the remaining parts, the whole system should be switched off immediately [24]. The output current paths under a switch short-circuit fault are shown in Fig. 2.15. Comparing Fig. 2.15 with the output current paths under normal operation (see Figs. 3.12 and 3.13), it can be seen that the output current path is changed when a short-circuit fault occurs. The output current paths in normal operation and short-circuit operation are denoted by the solid (in blue) and dotted (in red) lines, respectively. The short-circuited switching device is shown by a red solid line.

Fig. 2.15A shows the output current path for a short-circuit fault at S_{1k} under operating state 0 when $i_k > 0$. The short-circuit current flows through S_{3k} and D_{2k}. Clearly, the pole voltage which was zero before the short-circuit fault is not zero anymore. Fig. 2.15B shows the output current path for a short-circuit fault at S_{1k} under operating state N when $i_k < 0$. The short-circuit current flows through S_{4k}. In this case, the pole voltage which was $-V_s/2$ before the short-circuit fault is affected. Fig. 2.15C shows the output current path for a short-circuit fault at S_{2k} under operating state N when $i_k < 0$. The short-circuit current flows through D_{3k}. It is obvious that the inverter cannot produce a negative pole voltage.

2.2.4 Modulation of T-type inverter

While the switching combinations of a three-phase three-level T-type inverter yield 27 (3^3) voltage vectors, the single-phase three-level T-type inverter produces 9 (3^2) voltage vectors. Based on the magnitudes of the voltage vectors, they can be

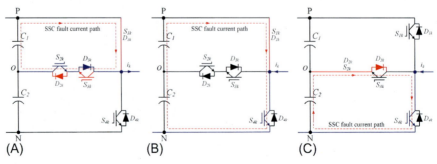

FIG. 2.15

Output current paths under a switch short-circuit fault: (A) $i_k > 0$ 0 state, (B) $i_k < 0$ N state, and (C) $i_k < 0$ N state.

represented by space voltage vectors which can be divided into six groups for the three-phase three-level T-type inverter and four groups for the single-phase three-level T-type inverter. The space voltage vectors of the three-phase three-level T-type inverter are depicted in Fig. 2.16. Based on the magnitudes of these vectors, there exist four groups which are called large, medium, small, and zero. While there are six vectors in the large (V_{13}, V_{14}, V_{15}, V_{16}, V_{17}, V_{18}) and medium (V_7, V_8, V_9, V_{10}, V_{11}, V_{12}) groups, there are twelve vectors in the small (V_1, V_2, V_3, V_4, V_5, V_6) group. The remaining three vectors are zero vectors (V_0). The lengths of the large, medium, and small vectors are $2V_s/3$, $\sqrt{3}V_s/3$, and $V_s/3$, respectively [21, 25, 26]. These space voltage vectors and their magnitudes, together with the switching states, are listed in Table 2.5. It is well known that space vector-based pulse width modulation (PWM) methods have attractive features such as good utilization dc-link voltage, low total harmonic distortion (THD), and easy implementation by using a digital signal processor [27].

It is also possible to produce PWM signals by using multicarrier methods which are based on traditional sinusoidal PWM with triangular carriers. One of the popular methods is to use level shifted carrier signals which are compared with a sinusoidal reference signal to generate PWM signals. In the case of a three-level T-type inverter, two triangular carrier signals are used. While the first carrier (Car1) changes between 0 and +1, the second carrier (Car2) changes between 0 and −1. The PWM modulation strategy and its block diagram for the three-level T-type inverter is shown

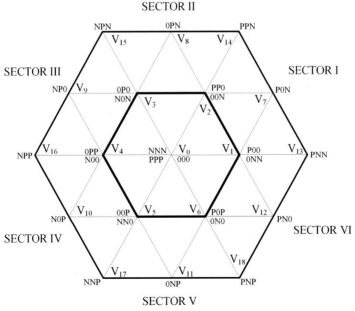

FIG. 2.16

Space voltage vectors of a three-level T-type inverter.

Table 2.5 Space voltage vectors and switching states.

Vector	Magnitude	Switching state
Zero (V_0)	0	[NNN] [000] [PPP]
Small ($V_1, V_2, V_3, V_4, V_5, V_6$)	$\frac{V_s}{3}$	P-type N-type [P00] [0NN] [PP0] [00N] [0P0] [N0N] [0PP] [N00] [P0P] [0N0] [00P] [NN0]
Medium ($V_7, V_8, V_9, V_{10}, V_{11}, V_{12}$)	$\frac{\sqrt{3}V_s}{3}$	[PON] [OPN] [NPO] [NOP] [ONP] [PNO]
Large ($V_{13}, V_{14}, V_{15}, V_{16}, V_{17}, V_{18}$)	$\frac{2V_s}{3}$	[PNN] [PPN] [NPN] [NPP] [NNP] [PNP]

in Fig. 2.17. Fig. 2.17A shows level shifted carrier signals together with a sinusoidal reference signal v_a^* used to produce the gate signals (g_{1a}, g_{2a}, g_{3a}, and g_{4a}) of the switching devices for phase a. The block diagram of the entire modulation strategy is depicted in Fig. 2.17B. The PWM based on multicarrier signals leads to a reduction in the switching frequency, which in turn results in a reduction in the switching losses.

2.2.5 Influence of the switching states on DC capacitor voltages

The switching states have some effects on the dc capacitor voltages V_{C1} and V_{C2} [21]. The relationship between the space voltage vectors and the switching state combinations is shown in Fig. 2.18. Fig. 2.18A shows the equivalent circuit for a large space vector [PNP]. Clearly, each leg is connected to a positive and negative dc link without having connection with the neutral point 0. In such a case, the capacitor voltages are not affected. The remaining large vectors also do not connect the inverter legs to the neutral point. Fig. 2.18B shows the equivalent circuit for the zero switching state [NNN]. Similarly, the neutral point is not connected to any of the inverter legs. For the switching state [PPP] (not shown), the neutral point is also not connected to any inverter legs. Although the neutral point is connected to all inverter legs in the switching state [000] (not shown), the capacitor voltages are not affected since the sum of the three-phase current is zero. Therefore, it can be concluded that large and zero vectors cannot influence the dc capacitor voltages V_{C1} and V_{C2}.

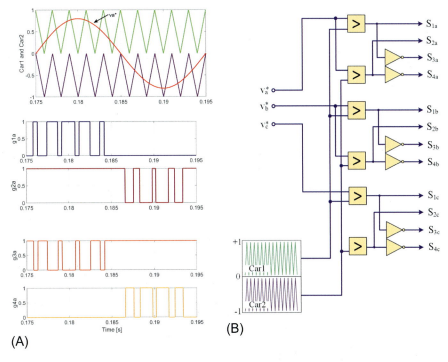

FIG. 2.17

PWM modulation strategy and its block diagram for a three-level T-type inverter: (A) PWM modulation strategy, and (B) block diagram.

Fig. 2.18C and D shows the equivalent circuits for the small switching states [PP0] and [00N], respectively. While [PP0] is a P-type switching state, [00 N] is an N-type switching state as shown in Table 2.5. In Fig. 2.18C, the neutral current i_0 flows into the neutral point since the upper dc capacitor C_1 is connected between the positive dc link and neutral point 0. In this case, while V_{C1} decreases, V_{C2} increases. On the other hand, in Fig. 2.18D, the neutral current i_0 flows from the neutral point since the lower dc capacitor C_2 is connected between the negative dc link and neutral point 0. In this case, the switching state decreases V_{C2} and increases V_{C1}. The remaining switching states in the small vector group also affect the dc capacitor voltages in the similar way.

Fig. 2.18E shows the equivalent circuit for the medium switching state [PN0]. Clearly, the neutral point is connected to inverter leg c while other legs are connected to the positive and negative dc link. In this case, the dc capacitor voltages increase and decrease depending on the direction of the neutral current i_0.

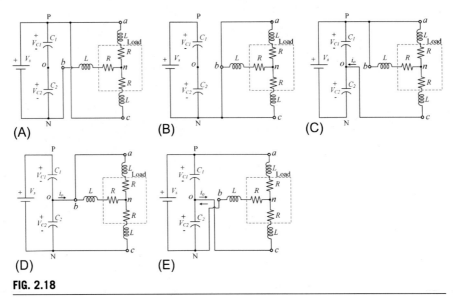

FIG. 2.18

Equivalent circuits for large, zero, small, and medium switching states: (A) [PNP] (large) state, (B) [NNN] (zero) state, (C) [PPO] (small) state, (D) [OO N] (small) state, and (E) [PNO] (medium) state.

2.2.6 Simulation results

In this section, simulation results are presented and discussed to validate the theoretical discussions in the previous sections regarding the switch open-circuit fault [21]. It is worth noting that the simulations are performed only considering a single-switch fault of leg a. Fig. 2.19 shows the waveforms of line-to-line voltage (v_{ab}), pole voltage (v_{a0}), and dc capacitor voltages (V_{C1} and V_{C2}) obtained under normal operation and open-circuit faulty operation occurring at $t = 0.3$ s in S_{1a}, S_{2a}, S_{3a}, and S_{4a}, respectively. In the simulation study, the system parameters are considered to be $V_s = 400$ V, $C_1 = C_2 = 5000$ μF, $L = 20$ mH, and $R = 40$ Ω.

The simulation results can be explained by considering the aforementioned statements in Section 3.2.2. Fig. 2.19A shows the simulated waveforms of v_{ab}, v_{a0}, V_{C1}, and V_{C2} obtained under an open-circuit fault in S_{1a}. As pointed out in Section 3.2.2, when an open-circuit fault occurs in S_{1a} while the inverter is working in the P state, the pole voltage v_{a0} is changed from +V_s/2 to 0 V. This fact is clearly shown in Fig. 2.19A. In such a case, the small voltage vectors of P state such as [PP0], [P00], [P0P], medium voltage vectors such as [P0N], [PN0], and large voltage vectors [PNN], [PPN], and [PNP] are impossible since the switching state P does not exist under an open-circuit fault of S_{1a}. As a result of this, V_{C1} becomes larger than V_{C2}. Fig. 2.19B shows the simulated waveforms of v_{ab}, v_{a0}, V_{C1}, and V_{C2} obtained under an open-circuit fault in S_{2a} while the inverter is working in the state 0. Again,

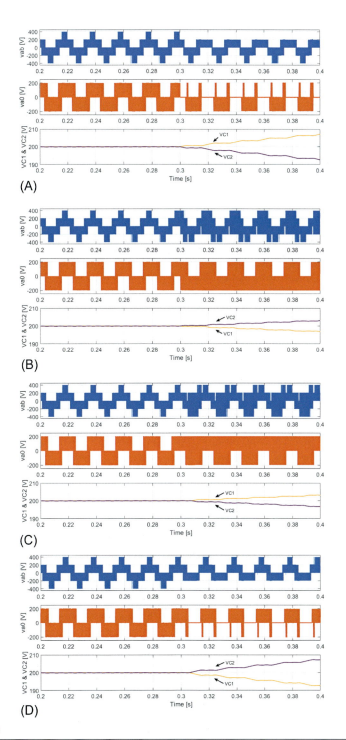

FIG. 2.19

Simulated waveforms of line-to-line voltage (v_{ab}), pole voltage (v_{a0}), and dc capacitor voltages (V_{C1} and V_{C2}) obtained under normal operation and open-circuit faulty operation occurring at $t = 0.3$ s in: (A) S_{1a}, (B) S_{2a}, (C) S_{3a}, and (D) S_{4a}.

as pointed out in Section 3.2.2, when an open-circuit fault occurs in S_{1a}, the pole voltage v_{a0} is changed from 0 V to $-V_s/2$ as shown in Fig. 2.19B. In this case, the small voltage vectors [0NN], [00 N], [0 N0] and medium voltage vectors [0PN] and [0NP] are impossible since the switching state 0 does not exist in the faulty leg. As a result of this, V_{C2} becomes larger than V_{C1}. Fig. 2.19C shows the simulated waveforms of v_{ab}, v_{a0}, V_{C1}, and V_{C2} obtained under an open-circuit fault in S_{3a} while the inverter is working in the state 0. As mentioned in Section 3.2.2, when an open-circuit fault occurs in S_{3a}, the pole voltage v_{a0} is changed from 0 V to $+V_s/2$ as shown in Fig. 2.19C. In this case, small voltage vectors [0P0], [0PP], [00P] and medium voltage vectors [0PN] and [0NP] are impossible since the switching state 0 does not exist in the faulty leg. As a result of this, V_{C1} becomes larger than V_{C2}. Finally, Fig. 2.19D shows the simulated waveforms of v_{ab}, v_{a0}, V_{C1}, and V_{C2} obtained under an open-circuit fault in S_{4a} while the inverter is working in the state N. As mentioned in Section 3.2.2, when an open-circuit fault occurs in S_{4a}, the pole voltage v_{a0} is changed from $-V_s/2$ to 0 V as shown in Fig. 2.19D. In this case, small voltage vectors [N0N], [N00], [NN0], medium voltage vectors [NP0] and [N0P], and large voltage vectors [NPN], [NPP], and [NNP] are impossible since the switching state N does not exist in the faulty leg. As a result of this, V_{C2} becomes larger than V_{C1}.

2.3 Conclusion

This chapter provides a comprehensive analysis on the three-level neutral-point-clamped (NPC) inverter and T-type inverter. A number of topics are described, including the inverter configuration, operating principle, switching states, switch fault analysis, influence of switching states on the dc capacitor voltages and modulation of the NPC and T-type inverter topologies. Simulation studies are also presented for NPC and T-type inverters.

The traditional NPC inverters and H-Bridge/NPC inverters have been dominantly used in practical applications, especially in the medium voltage drive industry. The power range varies from 0.15 to 72 MW; however, most of the installed MV drives in industry are in the range of 1–4 MW with voltage ratings of 3.3–6.6 kV.

The main disadvantage of the NPC inverter is the complexity of the power circuit and control due to the necessity of many active and passive components. For example, the floating capacitors add excessive cost to the overall system and require space for accommodating the entire volume. For this reason, the trend is to derive new topology from the NPC inverter that provides the same advantages but less passive and components counts.

Acknowledgment

The authors acknowledge the financial support from NPRP12S-0214-190083 from the Qatar National Research Fund (a member of the Qatar Foundation). The statements made herein are the sole responsibility of the authors.

References

[1] B. Wu, High-Power Converters and AC Drives, IEEE Press, A John Wiley & Sons. Inc., Publication, 2006.

[2] H. Abu-Rub, J. Holtz, J. Rodriguez, G. Baoming, Medium-voltage multilevel converters—State of the art, challenges, and requirements in industrial applications, IEEE Trans. Ind. Electron. 57 (8) (2010) 2581–2596.

[3] P. Barbosa, P. Steimer, J. Steinke, M. Winkelnkemper, N. Celanovic, in: Active-Neutral-Point-Clamped (ANPC) Multilevel Converter Technology, Proceedings of the European Conference on Power Electronics and Applications, September 11–14, Dresden, Germany, 2005.

[4] P. Kakosimos, S. Bayhan, H. Abu-Rub, in: Single-Phase Cascaded H-Bridge Neutral-Point Clamped Inverter: A Comparison Between MPC and PI Control, Proceedings of the Forty First Annual Conference of the IEEE Industrial Electronics Society, October 23–27, Florence, Italy, 2016.

[5] S. Bayhan, M. Trabelsi, H. Abu-Rub, in: Model Predictive Control of Three-Phase Three-Level Neutral-Point-Clamped qZS Inverter, Proceedings of the sixth International Conference on Power Engineering, Energy and Electrical Drives, 29 June-01 July, Bydgoszcz, Poland, 2016.

[6] S. Bayhan, P. Kakosimos, H. Abu-Rub, J. Rodriguez, in: Model Predictive Control Of Five-Level Neutral-Point-Clamped/H-Bridge Quasi-Impedance Source Inverter, Proceedings of the Forty First Annual Conference of the IEEE Industrial Electronics Society, October 23–27, Florence, Italy, 2016.

[7] S. Bayhan, H. Komurcugil, A sliding-mode controlled single-phase grid-connected quasi-Z-source NPC inverter with double-line frequency ripple suppression, IEEE Access 7 (1) (2019) 160004–160016.

[8] H. Mhiesan, S.S. Lee, Y. Wei, A. Mantooth, in: A New Family of 7-Level Boost Active Neutral Point Clamped Inverter, Proceedings of the seventh IEEE Workshop on Wide Bandgap Power Devices and Applications (WiPDA), Raleigh, NC, USA, 2019, pp. 20–24.

[9] I. Colak, E. Kabalci, R. Bayindir, Review of multilevel voltage source inverter topologies and control schemes, Energy Convers. Manag. 52 (2) (2011) 1114–1128.

[10] A. Bellini, S. Bifaretti, Comparison Between Sinusoidal PWM and Space Vector Modulation Techniques for NPC Inverters, IEEE Russia Power Tech, 2005.

[11] S. Bayhan, H. Abu-Rub, Predictive Control of Power Electronic Converters, Power Electronics Handbook, Elsevier Science & Technology, 2017. ISBN13:9780128114070.

[12] S. Bayhan, M. Trabelsi, H. Abu-Rub, M. Malinowski, Finite control set model predictive control for a quasi-Z-source four-leg inverter under unbalanced load condition, IEEE Trans. Ind. Electron. 64 (4) (2017) 2560–2569.

[13] S. Bayhan, M. Mosa, H. Abu-Rub, Model predictive control of Z source inverters, in: Impedance Source Power Electronic Converters, IEEE & Wiley Press, 2016. ISBN: 9781119037071.

[14] R. Vargas, P. Cortes, U. Ammann, J. Rodriguez, J. Pontt, Predictive control of a three-phase neutral-point-clamped inverter, IEEE Trans. Ind. Electron. 54 (5) (2007) 2697–2705.

[15] M. Schweizer, J.W. Kolar, Design and implementation of a highly efficient three-level T-type converter for low-voltage applications, IEEE Trans. Power Electron. 28 (2) (2013) 899–907.

[16] J.W. Kolar, High efficiency drive system with 3-level T-type inverter, in: Proceedings of the fourteenth European Conference on Power Electronics and Applications, EPE11, Birmingham, United Kingdom, 2011, pp. 1–10. September 30–October 1.
[17] W. Zhang, C. Ding, Mitigation of the low-frequency neutral-point current for three-level T-type inverters in three-phase four-wire systems, IET Power Electron. 11 (8) (2018) 1–8.
[18] J. Chen, C. Zhang, A. Chen, X. Xing, Fault-tolerant control strategies for T-type three-level inverters considering neutral-point voltage oscillations, IEEE Trans. Ind. Electron. 66 (4) (2019) 2837–2846.
[19] U.M. Choi, K.B. Lee, Space vector modulation strategy for neutral-point voltage balancing in three-level inverter systems, IET Power Electron. 6 (7) (2013) 1390–1398.
[20] M. Schweizer, T. Friedli, J.W. Kolar, Comparative evaluation of advanced three-phase three-level inverter/converter topologies against two-level systems, IEEE Trans. Ind. Electron. 60 (12) (2013) 5515–5527.
[21] U.M. Choi, F. Blaabjerg, K.B. Lee, Reliability improvement of a T-type three-level inverter with fault-tolerant control strategy, IEEE Trans. Power Electron. 30 (5) (2015) 2660–2673.
[22] U.M. Choi, H.G. Jeong, K.B. Lee, F. Blaabjerg, Method for detecting an open-switch fault in a grid-connected NPC inverter system, IEEE Trans. Power Electron. 27 (6) (2012) 2726–2739.
[23] S. Xu, J. Zhang, J. Hang, Investigation of a fault-tolerant three-level T-type inverter system, IEEE Trans. Ind. Appl. 53 (5) (2017) 4613–4623.
[24] U.M. Choi, K.B. Lee, F. Blaabjerg, Diagnosis and tolerant strategy of an open-switch fault for T-type three-level inverter systems, IEEE Trans. Ind. Appl. 50 (1) (2014) 495–508.
[25] U.M. Choi, H.H. Lee, K.B. Lee, Simple neutral-point voltage control for three-level inverters using discontinuous pulse width modulation, IEEE Trans. Energy Convers. 28 (2) (2013) 434–443.
[26] C.R. Clemente, E.R. Cadaval, M.R. Cortes, O. Husev, Carrier level-shifted based control method for the PWM 3 L-T-type qZS inverter with capacitor imbalance compensation, IEEE Trans. Ind. Electron. 65 (10) (2018) 8297–8306.
[27] J. Rodriguez, J.S. Lai, F.Z. Peng, Multilevel inverters: a survey of topologies, controls, and applications, IEEE Trans. Ind. Electron. 49 (4) (2002) 724–738.

CHAPTER 3

Conventional H-bridge and recent multilevel inverter topologies

Ilhami Colak[a], Ersan Kabalcı[b], and Gokhan Keven[c]

[a]*Department of Electrical and Electronics Engineering, Faculty of Engineering and Architecture, Nisantasi University, Istanbul, Turkey* [b]*Department of Electrical and Electronics Engineering, Faculty of Engineering and Architecture, Nevsehir Haci Bektas Veli University, Nevsehir, Turkey* [c]*Department of Electronics and Automation, Vocational High School, Nevsehir Haci Bektas Veli University, Hacibektas, Nevsehir, Turkey*

3.1 Introduction

Today, generation of electricity from renewable energy sources (RESs) is a key issue in distributed energy generation systems. Photovoltaic (PV) panels, wind turbines, fuel cells, biomass, and other natural sources are used for energy production [1]. RESs help to protect the planet from pollution caused by conventional energy generation systems. In addition, RESs have another advantage in that the energy is produced close to where it is consumed. PVs became commercially viable at the beginning of this century, converting freely available sunlight into electricity cost-effectively [2]. The International Energy Agency (IEA) has reported that the total PV power plant capacity increased by more than 100 GW in 2019 [1, 2]. PV systems are classified into two categories: stand-alone and grid-tied PV systems. The stand-alone PV systems use the energy directly at load sites, but grid-tied systems are connected to the grid for transmission, distribution, and consumption. The percentage of energy produced by grid-tied systems is increasing daily [3].

The inverter converts the energy produced by PV panels from DC to AC. The connection between PV modules and the grid is made in two different ways, with galvanic isolation (with transformer or isolated) and without galvanic isolation (transformerless or nonisolated), as depicted in Fig. 3.1 [3, 4]. Galvanic isolation is provided by using a transformer in the inverter connected to the grid. This type of isolation protects the system and users from hazardous voltages and leakage currents. In addition to this, it reduces the interference of the leakage current and DC injection into the grid. Grid-tied PV systems with a transformer are implemented with a low-frequency (line frequency) or high-frequency transformer-based inverter, shown in Fig. 3.1A and B, respectively. The low-frequency transformers, which are heavy, bulky, and expensive solutions, are located at the AC side of

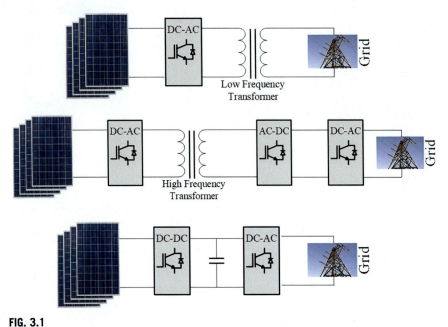

FIG. 3.1

Classification of grid-tied inverters regarding galvanic isolation: (A) low-frequency transformer-based inverter; (B) high-frequency transformer-based inverter; (C) transformerless inverter.

the inverter, while the high-frequency transformer is used on the DC side of the inverter. It should be noted that low-frequency transformers reduce the system efficiency due to power loss in the windings. A crucial reduction in size and weight is obtained with a high-frequency transformer. However, the efficiency of the inverter is still low due to the conversion of energy between DC and AC more than once. Hence, nonisolated inverters are used in PV grid-tied systems owing to their high efficiency, lower cost, and high power density [2–6]. The nonisolated inverter systems do not have any galvanic isolation. Therefore leakage current is higher when compared with the isolated inverters. The leakage current increases the electromagnetic interference (EMI), total harmonic distortion (THD), and the system losses, in addition to safety problems caused by the leakage current [4].

PV inverters are commonly implemented in the H-bridge topology in both isolated and nonisolated systems. The H-bridge topology has four switching components in its traditional structure, which is called H4 topology, and it is not suitable for leakage current for nonisolated inverters. To achieve minimum leakage current, recent improved inverter topologies have emerged, such as H5, the highly efficient and reliable inverter concept (HERIC), and H6 configurations, with additional switching components in their structures [1]. Besides the structure of the inverter, modulation strategies are a very important issue in decreasing the leakage current and improving the output parameters of an inverter. Sinusoidal pulse width modulation (SPWM) is a common modulation strategy widely used in control of inverters [5].

3.2 H-bridge inverter topology

H-bridge (or full-bridge) inverter topology was patented by Baker et al. [7]. This topology can be used as inverter cells in cascaded multilevel inverters. Other traditional inverter topologies, namely neutral-point clamped and flying capacitor clamped, were invented after the H-bridge topology. As mentioned earlier, the H-bridge topology is typically used as an H4 topology, especially in grid-tied inverter systems where four switches are formed with two legs and each leg has two switches serially connected, as seen in Fig. 3.2A. S_1-S_2 complementary switches

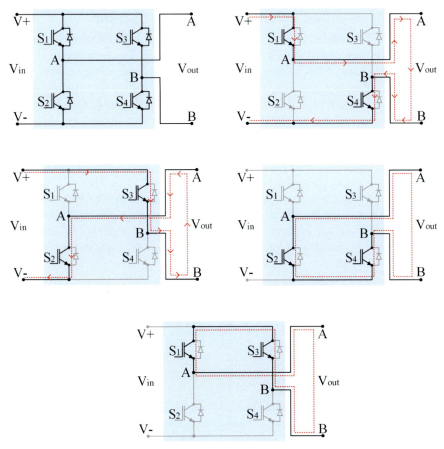

FIG. 3.2

H4 inverter: (A) general structure; (B) positive half-cycle; (C) negative half-cycle; (D) freewheeling mode-I; (E) freewheeling mode-II.

are connected on the same leg, while S_3-S_4 complementary switches are on the other leg. The positive and negative input terminals are connected to batteries or power supplies, and AC output terminals are connected to the middle point of the legs [8, 9].

This inverter generates three different output values, $+V_{DC}, 0, -V_{DC}$, for V_{DC} as the input value. The switching orders of the inverter are represented in Fig. 3.2B and C as the "ON" state of S_1 and S_4 switches in the positive half-cycle, and the "ON" state of S_2 and S_3 switches in the negative half-cycle of the AC wave. The zero-voltage state can be achieved in two different freewheeling methods. The S_2 and S_4 switches are turned to the "ON" state in the freewheeling mode-I, as shown in Fig. 3.2D, while the second one is obtained by turning S_1 and S_3 switches "ON" in freewheeling mode-II, as shown in Fig. 3.2E. The output value is "0" in these freewheeling modes [8, 9].

3.3 Common mode voltage and leakage current

H4 and recent H multilevel inverter topologies have been investigated related to analysis parameters in nonisolated multilevel PV inverter systems. The performances of the topologies are compared using parameters such as common mode voltage (CMV), leakage current, efficiency, and THD ratios. The efficiency denotes the conversion ratio between input and output power of the inverter, and the THD parameter is researched with regard to increasing the waveform quality of the inverter and grid [4, 10].

The CMV can be described as the average value of the voltages between the input or output terminals and a common reference. The negative terminal of the inverter (or PV panel) is the reference point (N) of the inverter output. Therefore the CMV of output terminals A and B can be calculated as shown in Eq. (3.1), where V_{AN} and V_{BN} are the voltage values between output terminals (A and B) of the inverter, and N is the reference point of the PV panel. The general diagram for a grid-tied nonisolated inverter with parasitic capacitance for the H4 inverter topology is shown in Fig. 3.3. The Z_G is the ground resistance of grid and parasitic capacitance connected

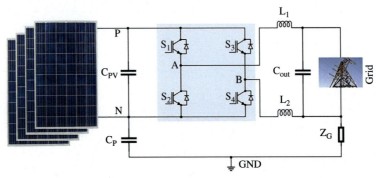

FIG. 3.3

General diagram for a grid-tied nonisolated inverter with parasitic capacitance.

via this resistor to grid. L_1-L_2 and C_{out} are filter components. L_1 and L_2 must have the same value to achieve minimum alternation in CMV [2, 6, 10].

$$V_{CM} = \frac{V_{AN} + V_{BN}}{2} \quad (3.1)$$

The leakage current is caused by CMV and it is detected from parasitic capacitance (C_P). The C_P occurs between PV modules and the grid as an unwanted side effect. An electrically chargeable surface area is accrued in the PV module that looks like a grounded frame. Therefore a connection occurs between the ground of the grid and PV module via the AC and DC filter elements and the grid impedance in a non-isolated inverter. The effect of C_P depends on many factors, such as the structure of the solar panel and the structure of the frame, surface of the cell, the distance between cells in the PV module, weather conditions, humidity, dust or salt covering the PV panels, etc. Therefore the C_P changes for each PV module under different conditions [4, 6, 10].

A simplified common mode leakage current model can be obtained by replacing the PV panel and switches with two PWM voltage sources (V_{AN} and V_{BN}), as shown in Fig. 3.4. The total leakage current (I_L) can be calculated by using the superposition theorem, which sums the currents in Eq. (3.2). Here, I_{L1}, I_{L2}, and I_{LG} are the leakage currents generated by V_{AN}, V_{BN}, and grid voltage as expressed in Eqs. (3.3)–(3.5), respectively. In these equations, L_1 and L_2 inductors are at the same value and they are used as L. The parameter I_{LG} can be neglected at grid frequency, because the

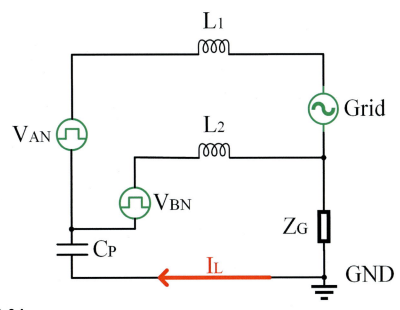

FIG. 3.4
Simplified common mode leakage current model.

leakage current depends on frequency. Therefore Eq. (3.2) can be written as Eq. (3.6). The total leakage current can be calculated in terms of CMV using Eq. (3.7). The equivalent circuits of leakage current (generated by V_{AN}, V_{BN}, and grid) are shown in Fig. 3.5A–C, respectively [2, 10].

$$I_L = I_{L1} + I_{L2} + I_{LG} \tag{3.2}$$

$$I_{L1} = \frac{jV_{AN}}{\frac{2}{\omega C_P} - \omega L} \tag{3.3}$$

$$I_{L1} = \frac{jV_{BN}}{\frac{2}{\omega C_P} - \omega L} \tag{3.4}$$

$$I_{LG} = \frac{-jV_G}{\frac{2}{\omega_g C_P} - \omega_g L} \tag{3.5}$$

$$I_L = \frac{jV_{AN}}{\frac{2}{\omega C_P} - \omega L} + \frac{jV_{BN}}{\frac{2}{\omega C_P} - \omega L} \tag{3.6}$$

$$I_L = \frac{j2V_{CMV}}{\frac{2}{\omega_{CMV} C_P} - \omega_{CMV} L} \tag{3.7}$$

As discussed earlier, the leakage current is generated due to the alternation of the CMV of the inverter. The minimum value of the variation of CMV causes minimum leakage current. Alternation of CMV decreases with the use of bipolar SPWM as a modulation strategy. In this situation, THD increases and the efficiency of the inverter decreases. When the inverter is switched using unipolar SPWM as the

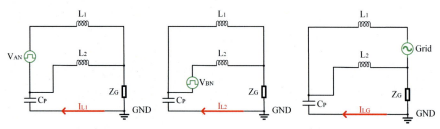

FIG. 3.5

The equivalent circuits of leakage current: (A) generated by V_{AN}, (B) generated by V_{BN}, (C) generated by grid.

Table 3.1 Leakage current values and their corresponding disconnection times.

Leakage current value (mA)	Disconnection time (s)
30	0.3
60	0.15
100	0.04

modulation strategy, alternation of CMV and THD ratios decreases, and the efficiency of the inverter increases. The H4 inverter topology is suitable to be used with two different SPWM modulation strategies, which are introduced in the following section of this chapter [11, 12]. The nonisolated grid-tied inverter systems must conform to some specific standards, such as IEEE 1547.1-2018, VDE0126-1-1, EN 50106, and IEC61727. The VDE0126-1-1 standard limits the RMS value of leakage current to 300 mA. Based on this standard, leakage current values and their corresponding disconnection times are listed in Table 3.1 [2, 13, 14].

3.4 Modulation strategy

Modulation strategy is one of the major factors of the nonisolated grid-tied H-bridge inverter. SPWM is the most commonly used modulation technique for H4 and other H-bridge based multilevel topologies. SPWM is a fundamental modulation method for generating with digital signal processors and it is convenient for decreasing leakage current, alternating CMV and THD of the inverter. There are three fundamental SPWM modulations applied in H4 inverter topology: bipolar SPWM, unipolar SPWM, and hybrid SPWM. The main differences among these modulation strategies are the switching patterns applied to the switching devices. A sinusoidal modulation signal is compared with a triangle wave carrier signal in the basic principle of the SPWM technique. This operation generates the switching signals required for switching the H-bridge semiconductors. These three strategies are implemented for an H-bridge inverter, as seen in Fig. 3.6. S_1 and S_2 are connected to the same leg, while S_3-S_4 devices are on the next leg. S_1 and S_3 semiconductors are called upper switches, and S_2 and S_4 semiconductors are called lower switches in this configuration.

3.4.1 Bipolar SPWM

Semiconductors in one leg are controlled as complementary in bipolar SPWM for preventing a short circuit of the DC power supply. While S_1 and S_4 are switched to the "ON" state, S_2 and S_3 must be in the "OFF" state. To achieve this switching pattern, a sinusoidal signal must be compared with only one triangle wave signal. The bipolar SPWM modulation strategy is also known as a two-level modulation,

CHAPTER 3 Conventional H-bridge

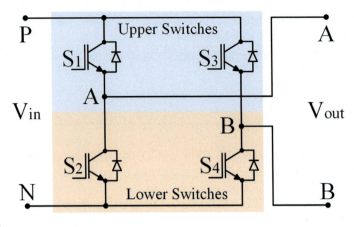

FIG. 3.6

The basic structure of H4 inverter.

which cannot provide a zero-voltage state at the output. If a DC source or a PV panel is connected as a V_{DC} source at the input, the output of the inverter will be changed between $+V_{DC}$ and $-V_{DC}$ when S_1-S_4 and S_2-S_3 are controlled at the same frequency. In this situation, the CMV (V_{AN} and V_{BN} values) becomes a constant value at $V_{DC}/2$ at the output of the inverter, which decreases the leakage current [15–18].

The advantages of bipolar SPWM modulation are stated as low EMI and low leakage current values; however, this method suffers from low efficiency, the requirement of high value filtering components, higher THD ratios, and higher core losses. Hence, the bipolar SPWM modulation strategy can be used in nonisolated grid-tied H4 inverters, but low efficiency and higher THD must be taken into consideration [15–18]. The generation of bipolar SPWM is illustrated in Fig. 3.7, where a sinusoidal modulation signal compared to a carrier triangle signal with "+1 V" and "−1 V" peak references. The generated switching orders seen in the second and third axes indicate that S_1 and S_4 switches are turned "ON" while S_2 and S_3 switches are turned "OFF." The conduction states of switches in the inverter topology are shown in Fig. 3.8A for the first switching order. While S_2 and S_3 switches are turned "ON" in the second order, S_1 and S_4 switches are turned "OFF" as depicted in the last axis of Fig. 3.7, and the circuit configuration is shown in Fig. 3.8B. Table 3.2 shows the summary of switching states for bipolar SPWM. In the bipolar modulation strategy, there is no freewheeling mode in the H4 inverter. Hence, positive or negative input terminals of the inverter are not connected to the grid without connection of a DC source. The leakage current is not affected via a parasitic capacitor to grid neutral. This operation reduces leakage current but varying output voltage without any freewheeling mode causes an incremented THD value and lower efficiency [15–19].

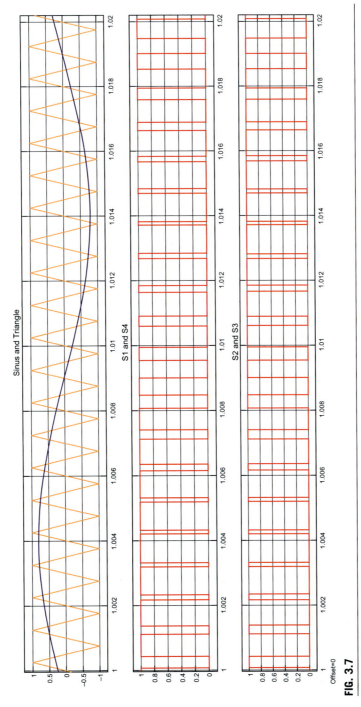

FIG. 3.7

Bipolar SPWM modulation strategy switching states.

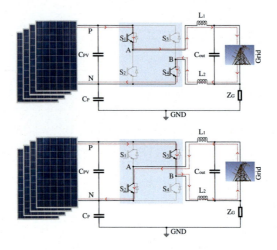

FIG. 3.8

Bipolar SPWM: (A) positive half-cycle (S_1-S_4 = "ON", S_2-S_3 = "OFF"), (B) negative half-cycle (S_2-S_3 = "ON", S_1-S_4 = "OFF").

Table 3.2 Switching states for bipolar SPWM.

Switching state	"ON" state switches	"OFF" state switches	V_{out}
Positive half-cycle	S_1-S_4	S_2-S_3	$+V_{DC}$
Negative half-cycle	S_2-S_3	S_1-S_4	$-V_{DC}$

3.4.2 Unipolar SPWM

The unipolar SPWM modulation, also known as three-level modulation, is used in H4 inverters instead of bipolar SPWM for reducing the THD value. The switching states are complementary in one branch of the H4 inverter. For example, S_1 and S_2 are switched in complementary states to each other, and S_3 and S_4 are also complementary to each other, as seen in Fig. 3.9. Two sinusoidal signals that are 180° phase shifted from each other are compared with a triangle carrier signal varying from "0" to "+1" for generating switching signals. If a DC source or a PV panel is connected as V_{DC} in the inverter input, the output levels of the inverter vary between $+V_{DC}$, 0, and $-V_{DC}$ [15–18]. The advantages of unipolar SPWM include decreased values for filtering the inverter output, lower core loss, and higher efficiency (up to 98%) due to reduced losses during the zero voltage state. The disadvantage of unipolar SPWM is that the EMI and leakage current are very high [18].

The generation of unipolar switching states is shown in Fig. 3.9. While the S_4 switching signal is at "ON" in the positive half-cycle, S_1 and S_2 are pulsed as complementary to each other at the carrier frequency. On the other hand, while the S_2 switching signal is at the "ON" position in the negative half-wave, S_3 and S_4 are

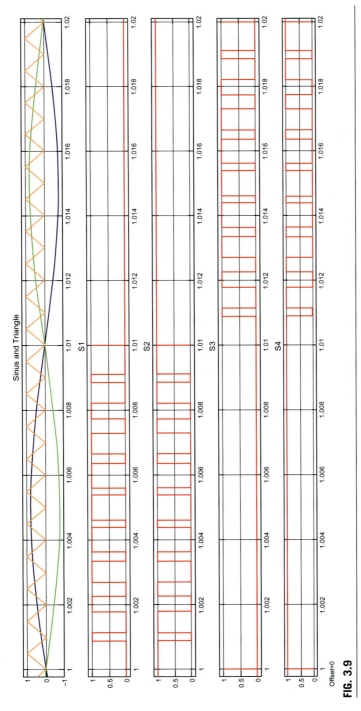

FIG. 3.9

Unipolar SPWM modulation strategy switching states.

pulsed as complementary to each other at the carrier frequency. There are four modes in the unipolar SPWM modulation strategy, as shown in Fig. 3.10. The positive half-cycle and negative half-cycle are depicted in Fig. 3.10A and B, respectively. Freewheeling mode-I and freewheeling mode-II are indicated in Fig. 3.10C and D, respectively. Freewheeling modes achieve zero voltage at the output. Unipolar SPWM is not suitable for nonisolated grid-tied inverters due to the high leakage current and variable CMV [15–18].

Table 3.3 summarizes the switching states of unipolar SPWM where freewheeling mode-I and freewheeling mode-II are identical but switching orders of devices are at different modulation frequencies in these modes. S_4 is always in the "ON" position and S_3 is always in the "OFF" position in freewheeling mode-I. S_1 and S_2 are controlled with complementary PWM signals in carrier frequency. S_2 is at the "ON" position and S_1 is at the "OFF" position in freewheeling mode-II. S_3 and S_4 are switched with complementary PWM signals in carrier frequency [19].

3.4.3 Hybrid SPWM

The hybrid SPWM method is another strategy, like unipolar SPWM, that supports freewheeling modes. One leg is switched at the fundamental frequency while the other leg is switched at the carrier frequency in the hybrid SPWM strategy, as the switching signals are depicted in Fig. 3.11. Two sinusoidal signals that are 180° phase shifted from each other are compared with a triangle carrier signal varying from "0" to "+1" for generating switching signals [10, 18, 19].

The advantages of hybrid SPWM are its lower losses due to unipolar voltage variation ($+V_{DC}$, 0, and V_{DC}), and higher efficiency (up to 98%). The disadvantages of hybrid SPWM are high leakage current and EMI filtering requirements, owing to square wave operation at the fundamental frequency in C_P voltage. Therefore hybrid SPWM is not preferred for nonisolated grid-tied inverter applications, even though it has high efficiency [18].

The current flow operation modes in hybrid SPWM switching states are like unipolar SPWM as depicted in Table 3.4. Hence, hybrid SPWM uses current paths like unipolar SPWM, as depicted in Fig. 3.10, but there are some differences in the frequency of switching in hybrid SPWM. S_1 is kept at the "ON" state and S_2 is turned to the "OFF" state in positive half-cycle and freewheeling mode-II. Furthermore, S_4 and S_3 switches are complementary at the carrier frequency. S_4 is at the "ON" state in the positive half-cycle while S_3 is at the "ON" state in freewheeling mode-II. Also, S_2 is at the "ON" state and S_1 is turned to the "OFF" state in the negative half-cycle and freewheeling mode-I. Furthermore, S_4 and S_3 switches are complementary at the carrier frequency. S_3 is at the "ON" state in the negative half-cycle and S_4 is at the "ON" state in freewheeling mode-I [10, 18, 19]. A comparison of these modulation schemes is shown in Fig. 3.12 from the study of [10]. Characteristics of bipolar SPWM in an H4 inverter are depicted in Fig. 3.12A, while characteristics of unipolar SPWM and hybrid SPWM in an H4 inverter are depicted in Fig. 3.12B and C.

3.4 Modulation strategy 69

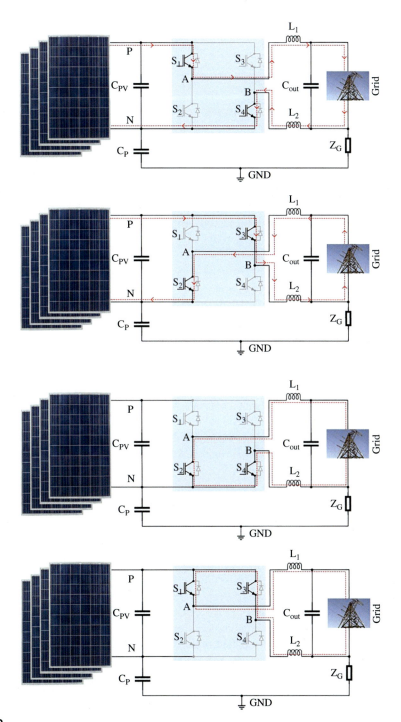

FIG. 3.10

Unipolar SPWM: (A) positive half-cycle, (B) negative half-cycle, (C) freewheeling mode-I, (D) freewheeling mode-II.

Table 3.3 Switching states for unipolar SPWM.

Switching State	"ON" state Switches	"OFF" state Switches	V_{out}
Positive half-cycle	S_1-S_4	S_2-S_3	$+V_{DC}$
Freewheeling mode-I	S_2-S_4	S_1-S_3	0
Freewheeling mode-II	S_1-S_3	S_2-S_4	0
Negative half-cycle	S_2-S_3	S_1-S_4	$-V_{DC}$

The output voltage and current waveforms of the different modulation strategies are fairly close to each other. Basic differences are seen in the variation of CMV caused by V_{AN} and V_{BN} values. The CMV value is constant in bipolar SPWM, as seen in Fig. 3.12A. Therefore the lowest leakage current (15.4 mA) is achieved in bipolar SPWM. The leakage current of unipolar SPWM is 1.8 A and the leakage current of hybrid SPWM is 3.9 A [10].

3.5 H5 inverter topology

H4 inverter topology is not quite suitable for the nonisolated grid-tied inverter, due to leakage current and CMV. The bipolar modulation strategy in the H4 inverter provides constant CMV and it enables low leakage current to be achieved, but the THD ratio and efficiency values of the H4 inverter are undesirable. The unipolar modulation strategy can be used for preventing the disadvantages of bipolar modulation, but the CMV does not present a constant value and the leakage current is increased. However, these limitations of the H4 inverter topology can be overcome by using an extra switch for cutting the connection of PV systems with the grid via the C_P capacitor. This topology is called H5 and it was proposed and patented by SMA Solar Technology in 2005 [17].

The H5 inverter topology has a different structure than the H4 topology, with an additional DC-bypass switch (S_5) that disconnects the PV from the grid in freewheeling modes and thus allowing constant CMV. Therefore, the leakage current is smaller than in H4 topology with the bipolar SPWM strategy. The H5 inverter topology, which is classified as a DC-bypass method, is depicted in Fig. 3.13 [20]. The switching states of H5 topology are listed in Table 3.5, with the positive half-cycle obtained while S_1, S_4, and S_5 switches are "ON" and S_2 and S_3 switches are "OFF", as seen in Fig. 3.14A. In the negative half-cycle, S_2, S_3, and S_5 switches are "ON" and S_1 and S_4 switches are "OFF", as seen in Fig. 3.14B. Two sinusoidal signals and one triangle signal are used for generating the switching order as in the unipolar SPWM. However, switching frequency applied to the switches is different from that in unipolar SPWM. The upper switches (S_1 and S_3) are switched at the fundamental frequency. However, the lower switches (S_2 and S_4) are switched at the carrier frequency (or higher frequency). The switching signals and switching orders are shown in Fig. 3.15 [4, 9, 11, 17–20].

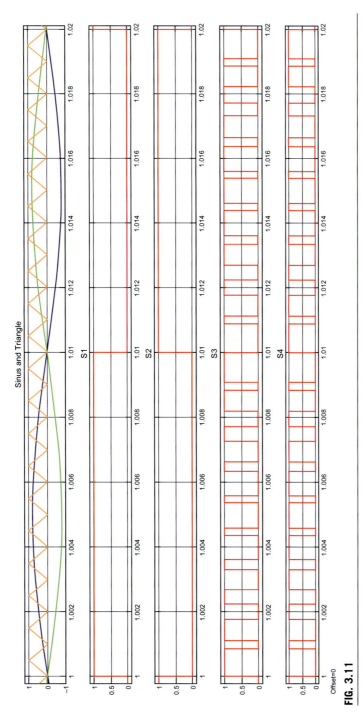

FIG. 3.11

Hybrid SPWM modulation strategy switching states.

Table 3.4 Switching states for hybrid SPWM.

Switching state	"ON" state switches	"OFF" state switches	V_{out}
Positive half-cycle	S_1-S_4	S_2-S_3	$+V_{DC}$
Freewheeling mode-I	S_2-S_4	S_1-S_3	0
Freewheeling mode-II	S_1-S_3	S_2-S_4	0
Negative half-cycle	S_2-S_3	S_1-S_4	$-V_{DC}$

There are two freewheeling modes available in the H5 inverter, as shown in Fig. 3.14C and D. Switching states of the freewheeling modes are shown in Table 3.5. All the switches are operated at the "OFF" state except the S_1 in freewheeling mode-I, but current flow in this mode is sustained over the antiparallel diode of S_3. Furthermore, all the switches are in the "OFF" state except the S_3 in freewheeling mode-II, so that the current flow is maintained over the antiparallel diode of S_1 in this mode. Also, the DC-bypass diode (S_5) is switched at the carrier frequency, and the S_5 switching signal is generated by the AND logic operation of S_2 and S_4. The advantages of the H5 inverter are lower core losses due to unipolar voltage variation, higher efficiency (up to 98%), fewer EMI filtering component requirements, and reduced leakage current. The disadvantages of the H5 inverter are higher conduction losses, but the overall high efficiency is not affected, and the addition of one extra switch [4, 9, 11, 17–20].

Characteristic output waveforms of the H5 inverter topology are shown in Fig. 3.16 from the study of [10]. V_{AB}, which is output voltage, varies between $+V_{DC}$, 0, $-V_{DC}$ and it decreases the CMV at the unipolar modulation. The variation of CMV is better than with unipolar SPWM and hybrid SPWM but is higher than the bipolar SPWM. Hence, the THD values of H5 topology (1%) are lower than with bipolar SPWM (1.8%) while the leakage current of bipolar SPWM (15.4 mA) is the lowest, compared to the leakage current of the H5 inverter at 23.4 mA [10].

3.6 H6 inverter topology

The H6 inverter, introduced by Ingeteam, is another DC bypass topology improved in order to reduce the CMV in inverters for achieving less leakage current. There are six switches in the H6 topology, as its name implies. An extra switch is connected in the negative bus of the DC-link of the H5 topology, as seen in Fig. 3.17. The switches connected to the positive and negative bus of the DC-link are used to disconnect the PV source from the grid in freewheeling mode. Therefore, C_P and its effect on CMV are decreased and inverter operation can be achieved with less leakage current [1, 2, 9, 10].

The switching states of the inverter are shown in Fig. 3.18, where it can be seen that there are two sinusoidal signals and one triangle signal is used in the comparator, as in unipolar SPWM. However, the H-bridge section of the H6 inverter topology

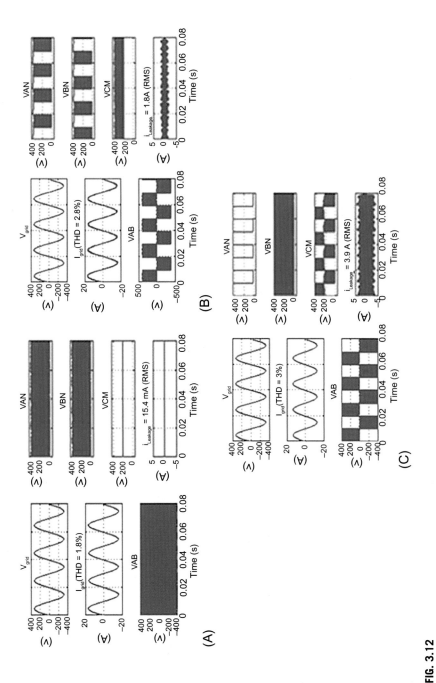

FIG. 3.12

Characteristic output waves of H4 inverter at different modulation schemes: (A) bipolar SPWM, (B) unipolar SPWM, (C) hybrid SPWM [10].

CHAPTER 3 Conventional H-bridge

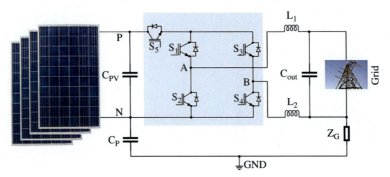

FIG. 3.13

H5 inverter topology.

Table 3.5 Switching states of H5 topology.

Switching state	"ON" state switches	"OFF" state switches	V_{out}
Positive half-cycle	S_1-S_4-S_5	S_2-S_3	$+V_{DC}$
Freewheeling mode-I	S_1-S_3(antiparallel diode)	S_2-S_3-S_4-S_5	0
Freewheeling mode-II	S_3-S_1(antiparallel diode)	S_1-S_2-S_4-S_5	0
Negative half-cycle	S_2-S_3-S_5	S_1-S_4	$-V_{DC}$

operates as in the hybrid SPWM strategy. The S_1 and S_2 switch legs are switched at the fundamental frequency, and S_3 and S_4 are switched at the carrier frequency. The S_5 switch is turned "ON" during the negative half-cycle that is switched at the carrier frequency by the logic AND operation of S_1 and S_4 in the positive half-cycle. The S_6 switch that is switched at the carrier frequency generated by the logic AND operation of S_2 and S_3 in the negative half-cycle is operated at the "ON" state during the positive half-cycle [2, 10].

The operation states of the topology are illustrated in Table 3.5, where the unipolar voltage variation ($+V_{DC}$, 0, $-V_{DC}$) occurs in the H6 topology. Switching states of the positive half-cycle given in Table 3.5 are depicted in Fig. 3.19A. Also, switching states of the negative half-cycle are shown in Fig. 3.19B. The freewheeling mode-I and freewheeling mode-II are illustrated in Fig. 3.19C and D, respectively.

Positive half-cycle states and freewheeling mode-I are successively operated in positive half-cycle, while the freewheeling mode-II is operated in negative half-cycle states (Table 3.6).

Characteristic output waves of the H6 inverter are shown in Fig. 3.20 [10]. V_{AB}, which is output voltage, varies between $+V_{DC}$, 0, $-V_{DC}$ and lower CMV is seen at this unipolar operation. The variation of CMV rates obtained in bipolar modulation is better than in unipolar and hybrid SPWM in the H6 topology. A small difference is noted between the leakage current of bipolar SPWM (15.4 mA) and the leakage

3.6 H6 inverter topology   **75**

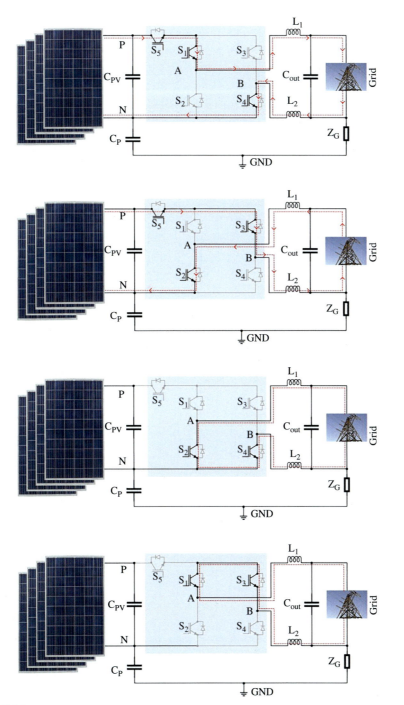

FIG. 3.14

H5 inverter: (A) positive half-cycle, (B) negative half-cycle, (C) freewheeling mode-I, (D) freewheeling mode-II.

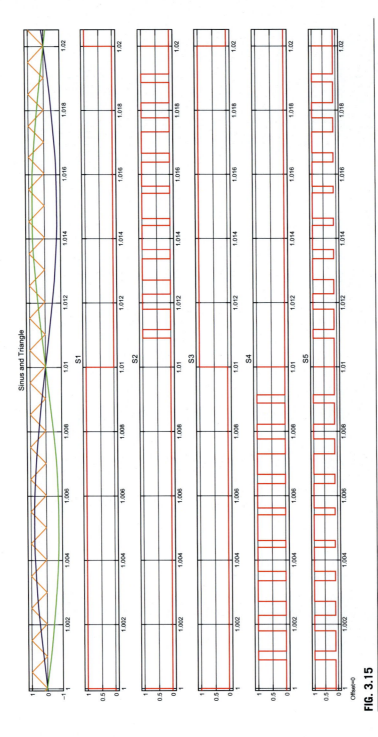

FIG. 3.15

Switching signals in H5 topology.

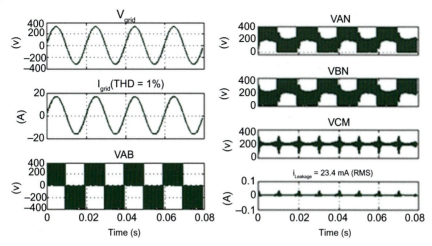

FIG. 3.16

Characteristic output waveforms of H5 inverter [10].

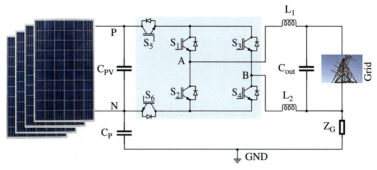

FIG. 3.17

H6 inverter topology.

current of the H6 inverter (16.09 mA), while the THD rates of H6 (1.1%) are better than with bipolar SPWM (1.8%); 1% was obtained with H5 inverter topology under similar operating conditions [10].

3.7 HERIC inverter

The HERIC topology was proposed by Schmidt et al. Two switches (S_A and S_B) are connected to the AC output of the H4 topology in the HERIC inverter, as shown in Fig. 3.21. The topology is implemented to decrease the leakage current and increase the efficiency by using a bypass operation at the AC side of the inverter. These

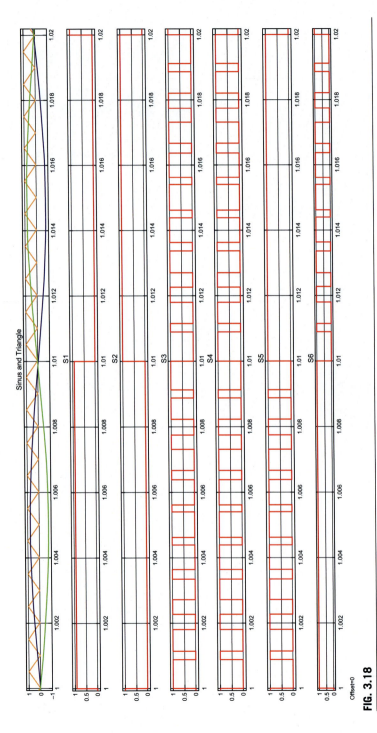

FIG. 3.18

Switching signals in H6 topology.

3.7 HERIC inverter

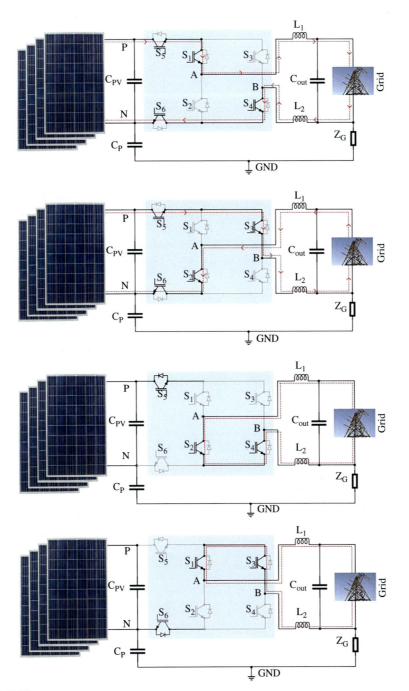

FIG. 3.19

H6 inverter: (A) positive half-cycle, (B) negative half-cycle, (C) freewheeling mode-I, (D) freewheeling mode-II.

CHAPTER 3 Conventional H-bridge

Table 3.6 Switching states of H6 topology.

Switching state	"ON" state switches	"OFF" state switches	V_{out}
Positive half-cycle	S_1-S_4-S_5-S_6	S_2-S_3	$+V_{DC}$
Freewheeling mode-I	S_1-S_3-S_6	S_2-S_4-S_5	0
Freewheeling mode-II	S_2-S_4-S_5	S_1-S_3-S_6	0
Negative half-cycle	S_2-S_3-S_5-S_6	S_1-S_4	$-V_{DC}$

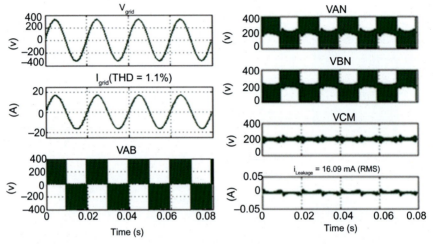

FIG. 3.20

Characteristic output waves of H6 inverter [10].

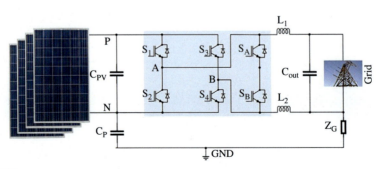

FIG. 3.21

HERIC inverter topology.

switches generate a zero voltage level in freewheeling modes. Also, the output of the inverter is disconnected from PV neutral in these operation modes. The HERIC inverter has three output voltage levels ($+V_{DC}$, 0, $-V_{DC}$) that are obtained by using a unipolar SPWM strategy. The switching order decreases the leakage current due to constant CMV. Furthermore, the efficiency of the inverter remains high because the output current is short-circuited via S_A or S_B during the freewheeling modes. The HERIC inverter is expected to perform well in nonisolated PV inverter systems, because of the high efficiency and very low leakage current and EMI [1, 2, 10, 14, 16].

The switching states of the HERIC inverter are shown in Fig. 3.22 and four operation modes are depicted in Table 3.7. The H-bridge switching devices (S_1, S_2, S_3, S_4) are switched at carrier (high) frequency and the AC bypass switches (S_A and S_B) are controlled at the fundamental frequency. In addition, the S_1-S_4 and S_2-S_3 switches are operated in a similar way to bipolar SPWM. The S_A and S_B are switched complementarily, and using AC bypass switches provides lower losses due to the reduced number of switches in the current conduction path [2, 10, 19, 21].

The positive half-cycle is illustrated in Fig. 3.23A. In this mode, S_1, S_4, and S_A are "ON" states where the S_1 and S_4 are switched at high frequency. While S_1 and S_4 are at the "OFF" state, the HERIC inverter operates in freewheeling mode-I, as seen in Fig. 3.23C. The current path crosses from S_A and the antiparallel diode of S_B in freewheeling mode-I. The S_2, S_3, and S_B are at "ON" state in the negative half-cycle, as depicted in Fig. 3.23B. While S_2 and S_3 are turned to the "OFF" state, the HERIC inverter operates in freewheeling mode-II, as seen in Fig. 3.23D. The current path crosses from S_B and the antiparallel diode of S_A in freewheeling mode-II [2, 10, 19, 21].

Characteristic output waveforms of the HERIC inverter are shown in Fig. 3.24 [10]. The current THD rises to 1.7% due to the modulation of H-bridge switches, but the output voltage varies between $+V_{DC}$, 0, $-V_{DC}$ and lower CMV is seen. The leakage current of the HERIC inverter is measured at 23.5 mA, which is lower than in unipolar SPWM (1.8 A) and hybrid SPWM (3.9 A) of the H4 inverter, while it is almost the same as that of the H5 inverter (23.4 mA); however, it is higher than in the H6 inverter (16.09 mA) and the bipolar SPWM of the H4 inverter (15.4 mA) [10].

3.8 Recent H-bridge based multilevel topologies

The topologies implemented for nonisolated inverters have been extensively studied by researchers. Basic topologies such as H4, H5, H6, and HERIC are still being developed for achieving constant CMV and reducing leakage current and THD ratios. Recent H-bridge topology based on nonisolated multilevel topologies includes additional diodes, switches, and capacitors in the structure. These extra components make the inverter more complicated and expensive.

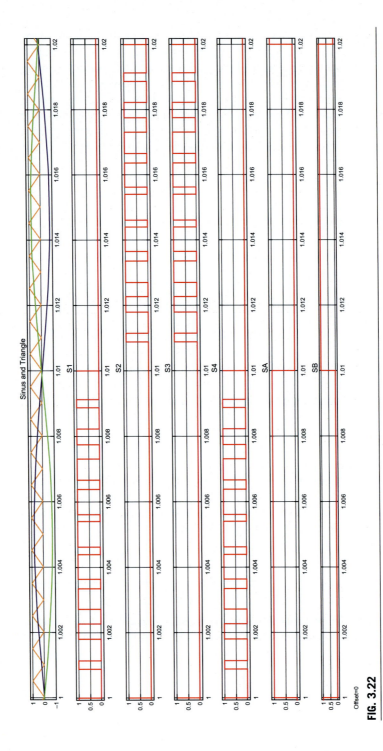

FIG. 3.22

Switching signals in HERIC topology.

Table 3.7 Switching states of HERIC topology.

Switching state	"ON" state switches	"OFF" state switches	V_{out}
Positive half-cycle	S_1-S_4-S_A	S_2-S_3	$+V_{DC}$
Freewheeling mode-I	S_A-S_B(antiparallel diode)	S_1-S_2-S_3-S_4	0
Freewheeling mode-II	S_B-S_A(antiparallel diode)	S_1-S_2-S_3-S_4	0
Negative half-cycle	S_2-S_3-S_B	S_1-S_4	$-V_{DC}$

3.8.1 Optimized H5 topology

The optimized H5 (oH5) topology has been proposed by Huafeng et al. in [22]. An extra switch is included as parallel to the positive terminal of the DC-link, as seen in Fig. 3.25. The sixth switch is connected between the middle point of the voltage divider capacitance and CMV is clamped to $V_{DC}/2$ in the freewheeling mode. The H-bridge switches of this inverter are controlled similarly to those of the hybrid SPWM. The switches are operated based on grid voltage polarity at high frequency or grid frequency, as depicted in Fig. 3.26. The switches of oH5, which are S_1-S_2, S_3-S_4, and S_5-S_6, are switched complementarily to each other and the operating frequency of the switches can be seen in Fig. 3.26. The switching state modes are represented in Table 3.8, and these states provide isolation between the grid and PV modules. S_5-S_6 are switched complementarily but they must have dead time in switching states to protect the capacitors from short circuits. Higher conduction losses are one of the other disadvantages of the oH5 inverter. Characteristic output waveforms of the HERIC inverter are shown in Fig. 3.27, which is presented in [10]. The lowest leakage current is achieved at 5.8 mA in the oH5 inverter and the THD of this inverter is measured at 1%. Hence, the oH5 inverter topology has proposed better power quality, comparing the discussed topologies [2, 10, 22].

3.8.2 H6-I and H6-II inverter topology

H6-I and H6-II inverters have six switches, as seen in Figs. 3.28 and 3.29, respectively. Two extra switches are added to the middle point of an H-bridge. The structure of these two topologies is similar. Differences in these topologies are the connection point of the diodes and the output of the inverter is connected to a different point of the inverter. Two freewheeling diodes are included in these inverters with the same connection method. The output of the inverter is connected to the upper side of the middle point switches (S_5 and S_6) in the H6-I topology. Also, the output of the inverter is connected to the lower side of the middle point switches (S_5 and S_6) in the H6-II topology. Advantages of H6-I and H6-II are high efficiency, low leakage current due to the parasitic capacitance, smaller output inductance, and no need for dead time [2, 5, 10, 23, 24].

Switching states are illustrated in Fig. 3.30 for the H6-I inverter topology and in Fig. 3.31 for the H6-II inverter topology. H-bridge switches (S_1, S_2, S_3, S_4) are switched in unipolar SPWM for the H6-I and H6-II inverters. Middle point switches

84 CHAPTER 3 Conventional H-bridge

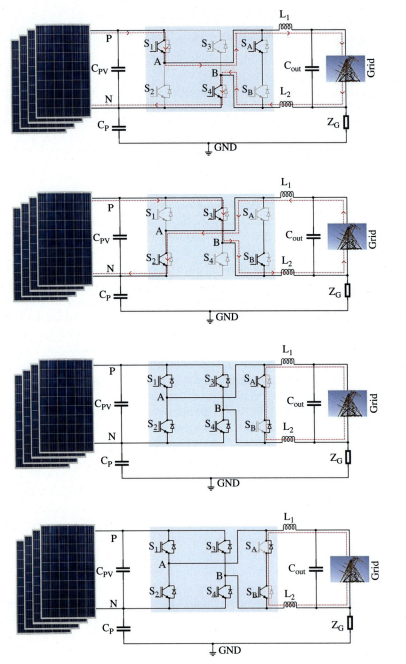

FIG. 3.23

HERIC inverter: (A) positive half-cycle, (B) negative half-cycle, (C) freewheeling mode-I, (D) freewheeling mode-II.

3.8 Recent H-bridge based multilevel topologies

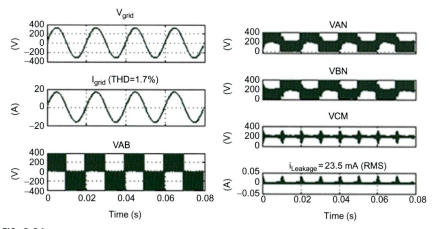

FIG. 3.24

Characteristic output waves of HERIC inverter [10].

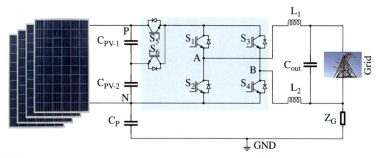

FIG. 3.25

The oH5 inverter topology.

(S_5 and S_6) are switched due to the polarity of output waves. S_5 is "OFF" and S_6 is "ON" for the positive half-cycle in H6-I, and vice versa in the negative half-cycle in H6-I inverter topology. The S_5 and S_6 switches operate inverse switching states from H6-I in the H6-II inverter topology [2, 5, 10, 23, 24].

The current path is explained in Tables 3.9 and 3.10 for the H6-I and H6-II inverter topologies, respectively. Only middle point switches change their position in the current path of positive, negative, and freewheeling modes. Characteristic output waves of these inverters are shown in Fig. 3.32. The leakage current is around 16 mA in H6-I and current THD is 2.6%, as depicted in Fig. 3.32A. Output values of the H6-II inverter topology are shown in Fig. 3.32B and here the leakage current is raised to 21.04 mA and THD is reduced to 2.3% [10].

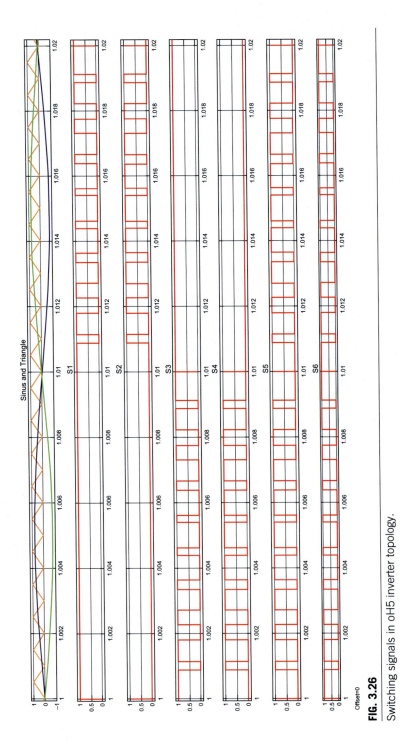

FIG. 3.26

Switching signals in oH5 inverter topology.

3.8 Recent H-bridge based multilevel topologies

Table 3.8 Switching states of oH5 topology.

Switching state	"ON" state switches	"OFF" state switches	V_{out}
Positive half-cycle	S_1-S_4-S_5	S_2-S_3-S_6	$+V_{DC}$
Freewheeling mode-I	S_1-S_3(antiparallel diode)	S_2-S_4-S_5-S_6	0
Freewheeling mode-II	S_3-S_1(antiparallel diode)	S_2-S_4-S_5-S_6	0
Negative half-cycle	S_2-S_3-S_5	S_1-S_4-S_6	$-V_{DC}$

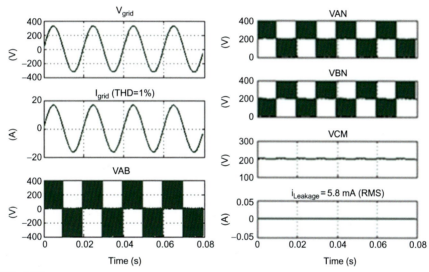

FIG. 3.27
Characteristic output waves of oH5 inverter [10].

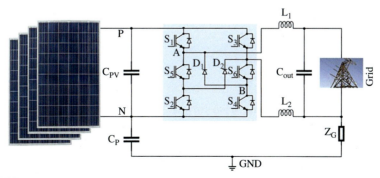

FIG. 3.28
H6-I inverter topology.

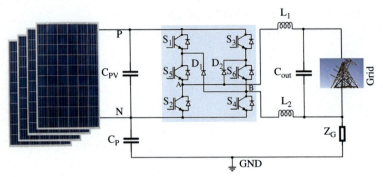

FIG. 3.29

H6-II inverter topology.

3.8.3 H6-III

The H6-III topology was proposed by Islam and Mekhilef in [5]. The structure of the H6-III topology was improved by adding a middle point, compared to H6-I and H6-II topologies. Two diodes acting as freewheeling modes in the middle points were removed. The structure of H6-III topology is depicted in Fig. 3.33, where the output of the inverter comprises from the upper point of the S_5 and S_6 switches. Unlike the H6-I and H6-II topologies, the H6-III topology is capable of injecting reactive power to the utility grid [5, 10]. The H-bridge switches, S_1-S_4, are switched at high frequency according to the polarity of the grid voltage, as shown in Fig. 3.34, while S_5 is in the "ON" state in positive half-cycle and is switched at high frequency in the negative half-cycle. S_6 is in the "ON" position in the negative half-wave and switches at high frequency in the positive half-wave.

While S_1-S_4 switches are in the "OFF" state, S_5 and S_6 switches provide the freewheeling path. Therefore, CMV is linear and does not alternate, as seen in Fig. 3.35. The current path of the H6-III inverter is explained in Table 3.11 for positive, negative, and freewheeling modes.

Lower leakage current is seen, at about 9.8 mA, and the THD ratio is remarkably low. Characteristic output waves of the H6-III inverter are illustrated in Fig. 3.35 [5, 10].

3.8.4 H6-IV topology

The H6-IV inverter topology, which includes six switches and two freewheeling diodes as shown in Fig. 3.36, was proposed by Cui et al. [25]. The S_1-S_4 switches are switched using hybrid SPWM in the following analyses. The S_1-S_2 switches are operated in fundamental frequency while the S_3-S_4 switches are triggered at the carrier frequency, as depicted in Fig. 3.37. In the positive half-cycle, S_1 is turned to the "ON" state and S_2, S_3, and S_6 are turned to the "OFF" state, while S_4 and S_5 are switched at the carrier frequency. In the negative half-cycle, S_3 and S_6 are switched at

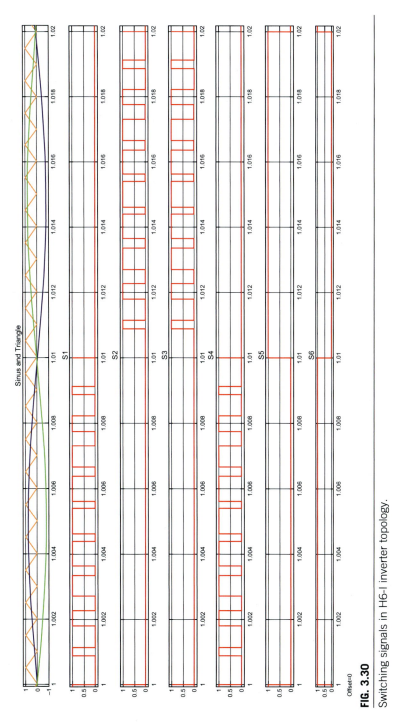

FIG. 3.30

Switching signals in H6-I inverter topology.

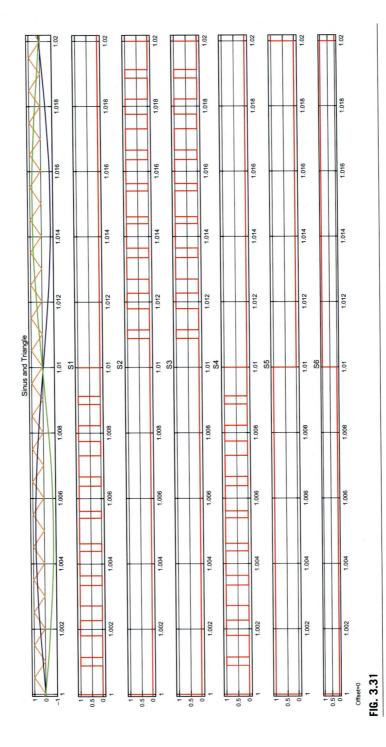

FIG. 3.31 Switching signals in H6-II inverter topology.

Table 3.9 Switching states of H6-I topology.

Switching state	"ON" state switches	"OFF" state switches	V_{out}
Positive half-cycle	S_1-S_4-S_6	S_2-S_3-S_5	$+V_{DC}$
Freewheeling mode-I	S_6-D_1	S_1-S_2-S_3-S_4-S_5	0
Freewheeling mode-II	S_5-D_2	S_1-S_2-S_3-S_4-S_6	0
Negative half-cycle	S_2-S_3-S_5	S_1-S_4-S_6	$-V_{DC}$

Table 3.10 Switching states of H6-II topology.

Switching state	"ON" state switches	"OFF" state switches	V_{out}
Positive half-cycle	S_1-S_4-S_5	S_2-S_3-S_6	$+V_{DC}$
Freewheeling mode-I	S_5-D_1	S_1-S_2-S_3-S_4-S_6	0
Freewheeling mode-II	S_6-D_2	S_1-S_2-S_3-S_4-S_5	0
Negative half-cycle	S_2-S_3-S_6	S_1-S_4-S_5	$-V_{DC}$

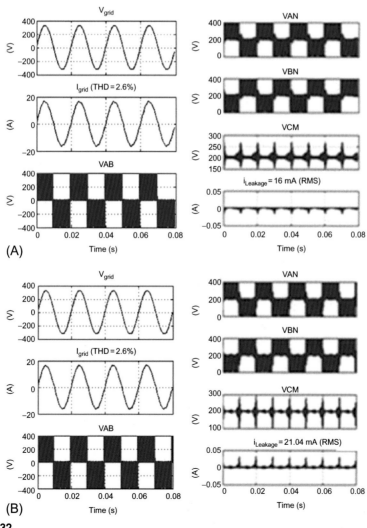

FIG. 3.32

Characteristic output waves of inverter: (A) H6-I topologies, (B) H6-II inverter topologies [10].

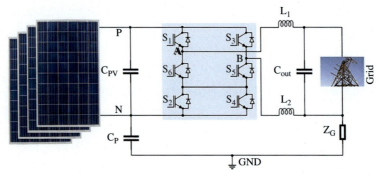

FIG. 3.33

H6-III inverter topology.

the carrier frequency, while S_2 is at the "ON" state and S_1, S_4, and S_5 are at the "OFF" state. The current path of freewheeling mode-I follows S_1-D_1 and the current path of freewheeling mode-II follows the S_2-D_2 path. The switching states of the modes are depicted in Table 3.12 [10, 25].

The simulation results of the H6-IV inverter are illustrated in Fig. 3.38, which has been represented in [10]. The CMV is not constant and oscillates during freewheeling periods; therefore the leakage current increases to 11 mA and it is not reduced completely [10].

3.8.5 Passive clamped H6 topology

The passive clamped H6 topology was proposed by Gonzalez et al. [26] as shown in Fig. 3.39. There are six switches (S_1-S_6) and two diodes (D_1 and D_2) in this structure. The diodes are connected to the middle point of voltage-divider capacitors. The connections of switching devices are configured similarly to the H6 inverter, but the switching states are different from the H6 inverter. The diodes D_1 and D_2 clamp the DC bus voltage to the V/2 level in the freewheeling mode. The switching states are illustrated in Fig. 3.40, where S5 and S6 switches are used to disconnect the grid from the PV in freewheeling mode [10, 26].

The operation of the topology in the power transfer stage and freewheeling modes is illustrated in Table 3.13. During the positive half-cycle, S_1 and S_4 are switched to the "ON" state, and S_5 and S6 are switched at the carrier frequency. Also, S_2 and S_3 are turned to the "ON" state, and S_5 and S_6 are switched at the carrier frequency in the negative half-cycle. The freewheeling modes of the passive clamped H6 topology are achieved at V/2 levels with clamping diodes and capacitors [10, 26].

The characteristic output waveforms of the passive clamped H6 inverter are depicted in Fig. 3.41 by simulation study of [10]. The inverter characteristics are improved due to clamping of CMV to the V/2 level, where the CMV is constant, unlike the H6 topology, and leakage current is very low, at 5.1 mA [10].

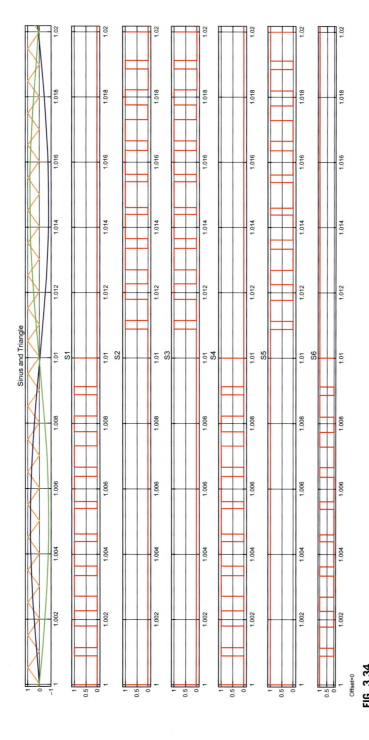

FIG. 3.34

Switching signals in H6-III inverter topology.

94 CHAPTER 3 Conventional H-bridge

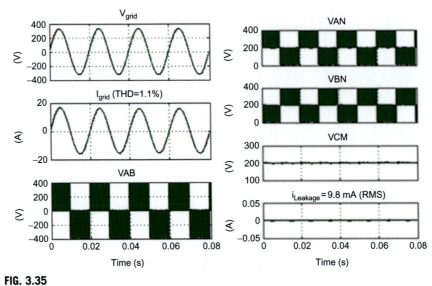

FIG. 3.35

Characteristic output waves of H6-III inverter [10].

Table 3.11 Switching states of H6-III topology.

Switching state	"ON" state switches	"OFF" state switches	V_{out}
Positive half-cycle	S_1-S_4-S_5	S_2-S_3-S_6	$+V_{DC}$
Freewheeling mode-I	S_5-S_6(antiparallel diode)	S_1-S_2-S_3-S_4	0
Freewheeling mode-II	S_6-S_5(antiparallel diode)	S_1-S_2-S_3-S_4	0
Negative half-cycle	S_2-S_3-S_5	S_1-S_4-S_6	$-V_{DC}$

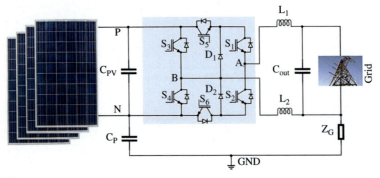

FIG. 3.36

H6-IV inverter topology.

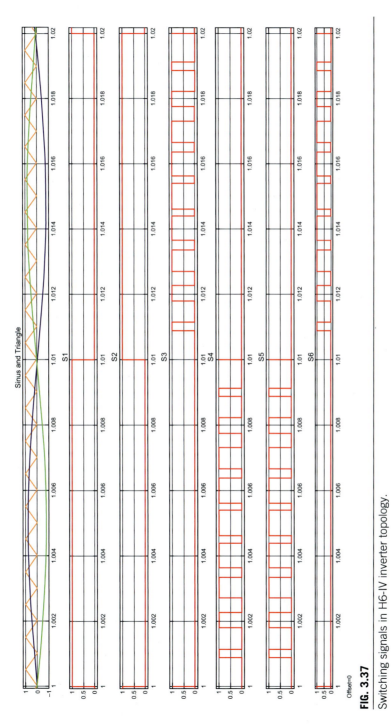

FIG. 3.37

Switching signals in H6-IV inverter topology.

96 CHAPTER 3 Conventional H-bridge

Table 3.12 Switching states of H6-IV topology.

Switching state	"ON" state switches	"OFF" state switches	V_{out}
Positive half-cycle	S_1-S_4-S_5	S_2-S_3-S_6	$+V_{DC}$
Freewheeling mode-I	S_1-D_1	S_2-S_3-S_4-S_5-S_6	0
Freewheeling mode-II	S_2-D_2	S_2-S_3-S_4-S_5-S_6	0
Negative half-cycle	S_2-S_3-S_6	S_1-S_4-S_5	$-V_{DC}$

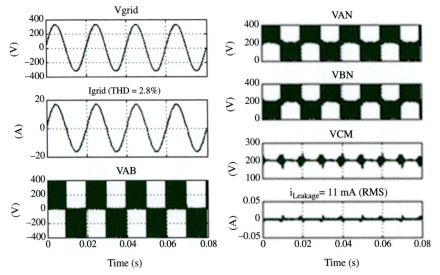

FIG. 3.38

Characteristic output waves of H6-IV inverter [10].

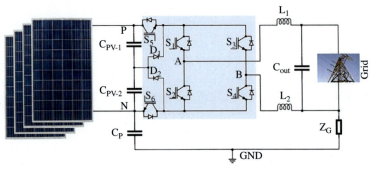

FIG. 3.39

Passive clamped H6 inverter topology.

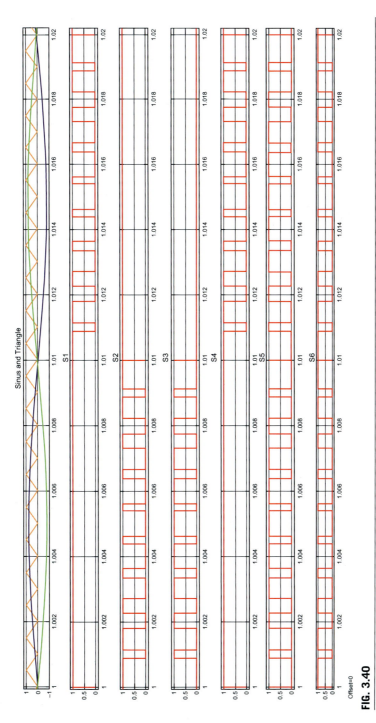

FIG. 3.40

Switching signals in passive clamped H6 inverter topology.

Table 3.13 Switching states of passive clamped H6 topology.

Switching state	"ON" state switches	"OFF" state switches	V_{out}
Positive half-wave	S_1-S_4-S_5-S_6	S_2-S_3	$+V_{DC}$
Freewheeling mode-I	S_1-S_3(antiparallel diode)	S_2-S_4-S_5-S_6	0
Freewheeling mode-II	S_3-S_1(antiparallel diode)	S_2-S_4-S_5-S_6	0
Negative half-wave	S_2-S_3-S_5-S_6	S_1-S_4-S_6	$-V_{DC}$

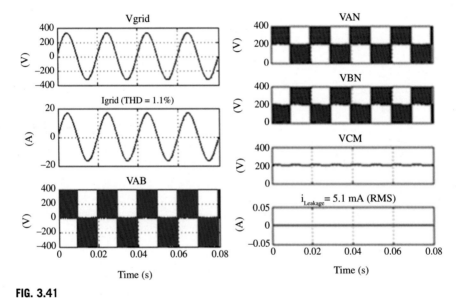

FIG. 3.41

Characteristic output waves of passive clamped H6 inverter [10].

3.8.6 HB-ZVR topology

The H-bridge zero voltage rectifier (HB-ZVR) topology was proposed by Kerekes et al. [27]. This topology provides constant CMV in freewheeling mode. The HB-ZVR topology is based on the H4 topology with an active switch (S_5), a full bridge rectifier (D_1-D_4), clamping capacitors (C_{PV-1} and C_{PV-2}) and a clamping diode (D_5) are included in the topology, as depicted in Fig. 3.42. This inverter topology forms an AC bypass with a rectifier bridge in freewheeling mode, similarly to the HERIC inverter [10, 27].

Operating states of the switching signal are shown in Fig. 3.43, and the current path of the HB-ZVR topology is depicted in Table 3.14. The inverter operates like an H4 topology in positive and negative half-cycles. The S_5 and rectifier bridge diodes operate in freewheeling mode. The S_5 switch functions complementarily for S_1 and

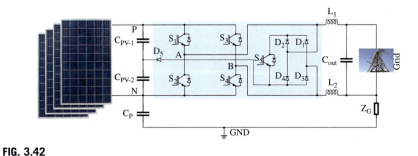

FIG. 3.42

HB-ZVR inverter topology.

S_4 in the positive half-cycle and S_2 and S_3 in the negative half-cycle. However, there must be dead time between H-bridge switches (S_1-S_4 or S_2-S_3) and the S_5 switch to protect the inverter from short circuits caused by clamping capacitors that are linear without fluctuation [10, 27].

Characteristic output waveforms of the HB-ZVR inverter are depicted in Fig. 3.44. HB-ZVR provides decreased leakage current compared to the HERIC inverter, which has an AC bypass structure. The leakage current of the HB-ZVR is 12.7 mA, while the THD ratio is the same as that of the HERIC inverter [10].

3.8.7 HBZVR-D topology

The H-bridge zero voltage rectifier-diode type topology (HBZVR-D) was proposed by Freddy et al. [21]. This topology is similar to the HBZVR, including an additional clamping diode in the structure connected to the middle point of the clamping diodes, as given in Fig. 3.45. Rectifier diodes comprise the zero-voltage state in freewheeling mode. The switching states are given in Fig. 3.46, where S_1, S_2, S_3, and S_4 switches play an active role in positive and negative half-cycle, and the devices are switched at the carrier frequency. S_5 is at the "ON" state while H-bridge switches are turned "OFF" in freewheeling modes.

The operations of switching states are depicted in Table 3.15. This is an improved version of HBZVR. The voltages of the D_{10} and D_{11} diodes clamp CMV to the $V_{DC}/2$ value in freewheeling modes. As a result, the leakage current is reduced to 5.4 mA and the THD ratio is detected as 1.2%, as illustrated in Fig. 3.47 [10, 21].

3.8.8 Active clamped HERIC topology

The active clamped HERIC configuration is an improved topology of the HERIC inverter from Li et al. [28]. This topology includes seven switches, where six (S_1, S_2, S_3, S_4 and S_A, S_B) are the same as in the HERIC inverter, as depicted in Fig. 3.48. An additional switch (S_C) is a clamping AC bypass switch proposed for clamping capacitors in freewheeling mode [10, 28].

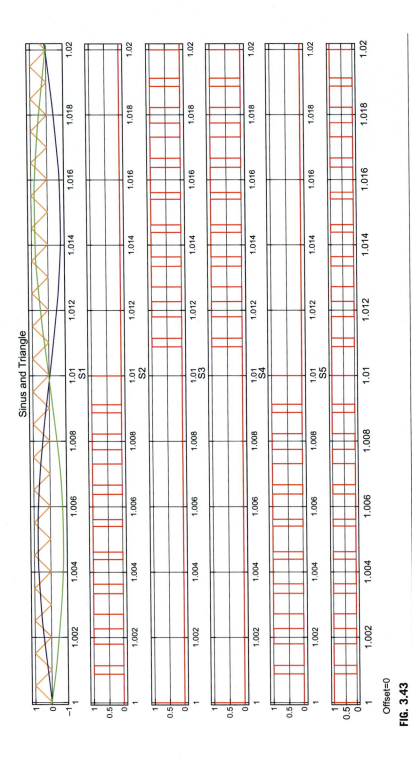

FIG. 3.43

Switching signals in HB-ZVR inverter topology.

3.8 Recent H-bridge based multilevel topologies

Table 3.14 Switching states of HB-ZVR topology.

Switching state	"ON" state switches	"OFF" state switches	V_{out}
Positive half-cycle	S_1-S_4	S_2-S_3	$+V_{DC}$
Freewheeling mode-I	S_5 and (D_1-D_3)	S_1-S_2-S_3-S_4	0
Freewheeling mode-II	S_5 and (D_2-D_4)	S_1-S_2-S_3-S_4	0
Negative half-cycle	S_2-S_3	S_1-S_4	$-V_{DC}$

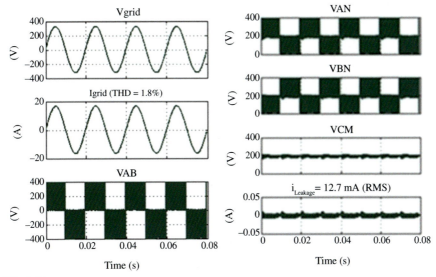

FIG. 3.44

Characteristic output waveforms of HB-ZVR inverter [10].

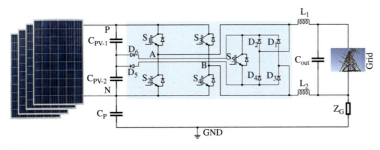

FIG. 3.45

HBZVR-D inverter topology.

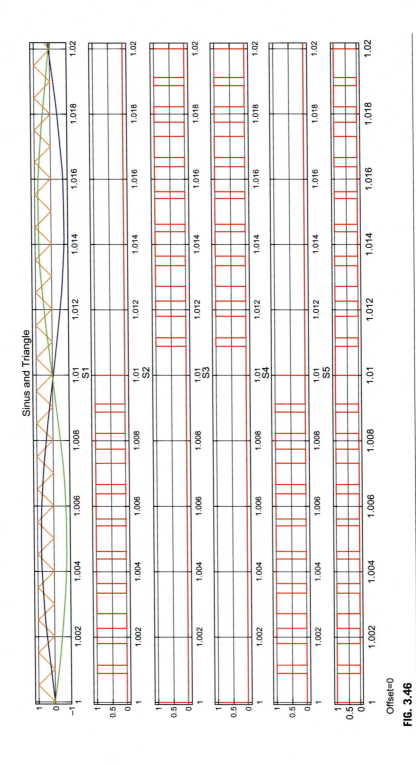

FIG. 3.46

Switching signals in HBZVR-D inverter topology.

3.8 Recent H-bridge based multilevel topologies

Table 3.15 Switching states of HBZVR-D topology.

Switching state	"ON" state switches	"OFF" state switches	V_{out}
Positive half-cycle	S_1-S_4	S_2-S_3	$+V_{DC}$
Freewheeling mode-I	S_5 and (D_1-D_3)	S_1-S_2-S_3-S_4	0
Freewheeling mode-II	S_5 and (D_2-D_4)	S_1-S_2-S_3-S_4	0
Negative half-cycle	S_2-S_3	S_1-S_4	$-V_{DC}$

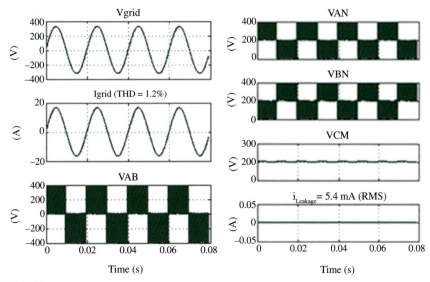

FIG. 3.47

Characteristic output waveforms of HBZVR-D inverter [10].

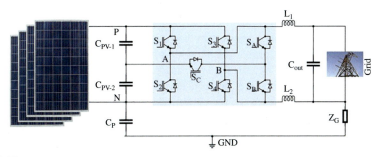

FIG. 3.48

Active clamped HERIC inverter topology.

The operating principle of the switches is illustrated in Fig. 3.49. The H-bridge switches are operated at carrier frequency as well as using the unipolar SPWM strategy. S_A and S_B are switched depending on the polarity of the output. In the positive half-cycle, S_A is "ON" and S_B is switched at carrier frequency as complementary of S_1 and S_4, while S_B is "ON" and S_A is switched at carrier frequency as complementary of S_2 and S_3 at the negative half-cycle. The freewheeling mode is given in Table 3.16. The switch S_C is not included in the freewheeling mode, but it is operated for clamping the output of the inverter. Characteristic output waveforms of active clamped HERIC inverter are shown in Fig. 3.50 [10].

3.9 Remarks and conclusion

The integration of multilevel inverters to distributed generation systems is implemented by using isolated or nonisolated topologies according to their galvanic isolation. Leakage current, efficiency, EMI, THD, and safety problems are commonly met with in nonisolated systems due to the absence of galvanic isolation. However, this issue can be tackled by using capacitive isolation.

Fundamental safety problems are caused by leakage current, which is produced by the fluctuation of CMV and parasitic capacitance between PV neutral and grid neutral, as handled in this chapter. Different SPWM strategies including bipolar, unipolar, and hybrid are used for switching classical H4 inverters. There are some advantages and disadvantages of these techniques. The output voltage of bipolar SPWM oscillates between $+V_{DC}$ and $-V_{DC}$ and these variations increase the dV/dt and THD ratios. However, no freewheeling period exists in the bipolar SPWM method and this provides minimum fluctuation of CMV and minimum leakage current. The unipolar and hybrid SPWM strategies oscillate between $+V_{DC}$, 0, and $-V_{DC}$.

Many new topologies are being proposed by researchers that are based on the H-bridge. The most common topologies are H5, H6, and HERIC. Also, there are recent H-bridge based topologies, such as oH5, H6-I, H6-II, H6-III, H6-IV, passive clamped H6, HB-ZVR, HBZVR-D and active clamped HERIC topologies. The purpose of these topologies is to reduce the leakage current and THD ratio. They are based on use of different numbers of switching components. Furthermore, some of them use extra diodes or capacitors in their structure. A comparison of topologies related to the component numbers is given in Table 3.17, where different component numbers are used in topologies due to the structure. The number of the switch is an important value in the structure, since if the number of switches is increased, the driven circuit number and cost of the system increases. Also, switches in the active current path (positive or negative half-cycle) are given in Table 3.17. When the number of active switch decreases, switching power loss of the system decreases. Topologies can be classified into two categories: DC bypass and AC bypass topologies. The important parameters of a nonisolated grid-tie PV inverter are leakage current and THD value. Leakage current values in different topologies are referred to the

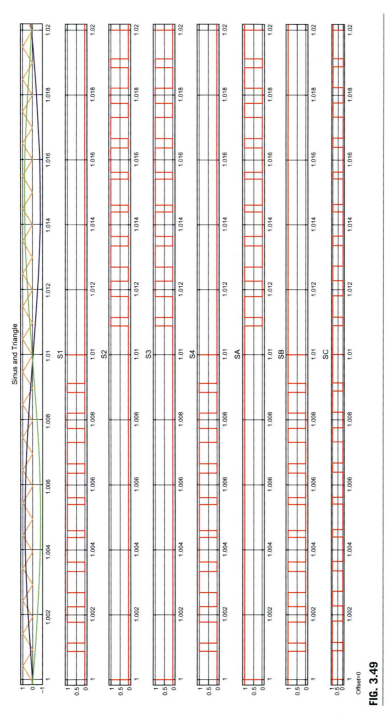

FIG. 3.49

Switching signals in active clamped HERIC inverter topology.

CHAPTER 3 Conventional H-bridge

Table 3.16 Switching states of active clamped HERIC topology.

Switching state	"ON" state switches	"OFF" state switches	V_{out}
Positive half-cycle	S_1-S_4	S_2-S_3-S_5-S_6	$+V_{DC}$
Freewheeling mode-I	S_A-S_B(antiparallel diode)	S_1-S_2-S_3-S_4	0
Freewheeling mode-II	S_B-S_A(antiparallel diode)	S_1-S_2-S_3-S_4	0
Negative half-cycle	S_2-S_3	S_1-S_4-S_5-S_6	$-V_{DC}$

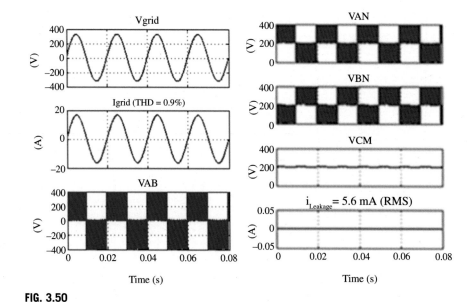

FIG. 3.50

Characteristic output waveforms of active clamped HERIC inverter [10].

simulations of [10], which are depicted in Fig. 3.51. The passive clamped H6 topology has the lowest leakage current values, but when comparing the total number of components and active switch in the current path with leakage current, the oH5 topology can be chosen due to its minimized device numbers and resultant efficiency. It is seen that the THD values of the topologies are close to each other and comply with the standards. The clamped CMV topologies provide reduced leakage current and THD value. The aim of these topologies is to produce constant CMV and minimum leakage current and THD values. Also, disconnection of PV panels from the grid neutral in freewheeling modes is allowed by these topologies.

Table 3.17 Comparison of nonisolated grid-tied H-bridge based topologies.

Inverter Topology	Total number of switches	Switches in the active current path	Switches in f path	Bypass mode	Leakage current	Input capacitor	THD (%)
H4 topology (Unipolar SPWM)	4S	2S	1 S	—	1.8 A	1	2.8
H4 topology (Hybrid SPWM)	4S	2S	1S + 1D	—	3.9 A	1	3
H4 topology (Bipolar SPWM)	4S	2S	1S + 1D	—	15.4 mA	1	1.8
H5 topology	5S	3S	1S + 1D	DC	23.4 mA	1	1
H6 topology	6S	4S	1S + 1D	DC	16.09 mA	1	1.1
HERIC topology	6S	2S	1S + 1D	AC	23.5 mA	1	1.7
H6-I topology	6S + 2D	3S	1S + 1D	DC	16 mA	1	2.6
H6-II topology	6S + 2D	3S	1S + 1D	DC	21.4 mA	1	2.3
H6-III topology	6S	3S	1S + 1D	DC	9.8 mA	1	1.1
H6-IV topology	6S + 2D	3S	1S + 1D	DC	11 mA	1	2.8
Optimized H5 topology	6S	3S	1S + 1D	DC	5.8 mA	2	1
Passive clamped H6 topology	6S + 2D	4S	1S + 1D	DC	5.1 mA	2	1.1
HB-ZVR topology	5S + 5D	2S	1S + 2D	AC	12.7 mA	2	1.8
ZVBR-D topology	5S + 6D	2S	1S + 2D	AC	5.4 mA	2	1.2
Active clamped HERIC topology	7S	2S	1S + 1D	AC	5.6 mA	2	0.9

S = Switch, D = Diode, f = Freewheeling mode.

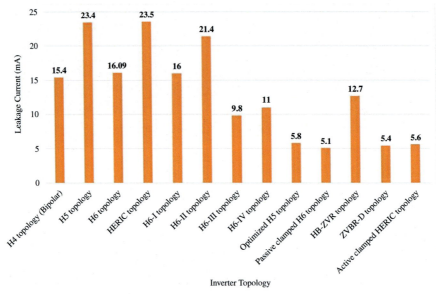

FIG. 3.51

Leakage current values in different topologies.

References

[1] E. Kabalcı, Review on novel single-phase grid-connected solar inverters: circuits and control methods, Solar Energy. 198 (2020) 247–274, https://doi.org/10.1016/j.solener.2020.01.063.

[2] M. Islam, S. Mekhilef, M. Hasan, Single phase transformerless inverter topologies for grid-tied photovoltaic system: a review, Renew. Sustain. Energy Rev. 45 (2015) 69–86, https://doi.org/10.1016/j.rser.2015.01.009.

[3] S. Chaudhary, S.N. Singh, in: Single phase grid interactive solar photovoltaic inverters: a review, 2018 National Power Engineering Conference, NPEC, 2018, 2018, pp. 1–6, https://doi.org/10.1109/NPEC.2018.8476767.

[4] A.A. Estévez-Bén, A. Alvarez-Diazcomas, G. Macias-Bobadilla, J. Rodríguez-Reséndiz, Leakage current reduction in single-phase grid-connected inverters—a review, Appl. Sci. 10 (2020) 2384. https://doi.org/10.3390/app10072384.

[5] M. Islam, S. Mekhilef, H6-type transformerless single-phase inverter for grid-tied photovoltaic system, IET Power Electron. 8 (2015) 636–644, https://doi.org/10.1049/iet-pel.2014.0251.

[6] N.M.J.S. Kirthiga, Highly reliable inverter topology with a novel soft computing technique to eliminate leakage current in grid-connected transformerless photovoltaic systems, Comput. Electr. Eng. 68 (2018) 192–203. https://doi.org/10.1016/j.compeleceng.2018.03.022.

[7] R.H. Baker, L.H. Bannister, Electric power converter, U.S. Patent No. 3,867,643. 18 Feb, 1975.

[8] M. Vijeh, M. Rezanejad, E. Samadaei, K. Bertilsson, A general review of multilevel inverters based on Main submodules: structural point of view, IEEE Trans. Power Electron. 34 (2019) 9479–9502, https://doi.org/10.1109/TPEL.2018.2890649.

[9] P.S. Gotekar, S.P. Muley, D.P. Kothari, B.S. Umre, Comparison of full bridge bipolar, H5, H6 and HERIC inverter for single phase photovoltaic systems—a review, 12th IEEE International Conference Electronics, Energy, Environment, Communication, Computer, Control: (E3-C3), INDICON 2015. (2016) 1–6. doi:https://doi.org/10.1109/INDICON.2015.7443837.

[10] Z. Ahmad, S.N. Singh, Comparative analysis of single phase transformerless inverter topologies for grid connected PV system, Sol. Energy 149 (2017) 245–271, https://doi.org/10.1016/j.solener.2017.03.080.

[11] H. Albalawi, S.A. Zaid, An H5 transformerless inverter for grid connected PV systems with improved utilization factor and a simple maximum power point algorithm, Energies 11 (2018) 1–17, https://doi.org/10.3390/en11112912.

[12] K. Chen, Z. Huang, L. Hang, X. Xie, Q. Han, Q. Lu, Y. Gan, G. Yang, P. Shen, Cascaded iH6 multilevel inverter with leakage current reduction for transformerless grid-connected photovoltaic system, in: Proceedings of the International Conference on Power Electronics and Drive Systems, 2017, pp. 613–617, https://doi.org/10.1109/PEDS.2017.8289270. Decem (2018).

[13] Z. Özkan, Leakage Current and Energy Efficiency Analyses of Single Phase Grid Connected Multi-kVA Transformerless Photovoltaic Inverters, Middle East Technical University, 2012.

[14] T. Salmi, M. Bouzguenda, A. Gastli, A. Masmoudi, A novel transformerless inverter topology without zero-crossing distortion, International Journal of Renewable Energy Research 2 (2012) 140–146, https://doi.org/10.20508/ijrer.03543.

[15] A. Namboodiri, H. Wani, Unipolar and bipolar PWM inverter, IJIRSRT - International Journal for Innovative Research in Science & Technology 1 (2014) 7. http://www.ijirst.org/articles/IJIRSTV1I7111.pdf.

[16] Y.R. Kafle, G.E. Town, X. Guochun, S. Gautam, Performance comparison of single-phase transformerless PV inverter systems, in: Conference Proceedings - IEEE Applied Power Electronics Conference and Exposition - APEC, 2017, pp. 3589–3593, https://doi.org/10.1109/APEC.2017.7931213.

[17] G. Rizzoli, M. Mengoni, L. Zarri, A. Tani, G. Serra, D. Casadei, Comparison of single-phase H4, H5, H6 inverters for transformerless photovoltaic applications, in: IECON Proceedings (Industrial Electronics Conference), 2016, pp. 3038–3045, https://doi.org/10.1109/IECON.2016.7792984.

[18] K. Zeb, I. Khan, W. Uddin, M.A. Khan, P. Sathishkumar, T.D.C. Busarello, I. Ahmad, H.J. Kim, A review on recent advances and future trends of transformerless inverter structures for single-phase grid-connected photovoltaic systems, Energies 11 (2018) 1–33, https://doi.org/10.3390/en11081968.

[19] M.I. Alzoubi, A.M.T. Oo, S. Azad, Simulation and evaluation of HERIC and H5 transformerless inverter topologies, in: *Proceedings of the 6th IASTED Asian Conference on Power and Energy Systems*, AsiaPES, 2013, 2013, pp. 212–219, https://doi.org/10.2316/P.2013.800-101.

[20] A.J. Babqi, Z. Yi, D. Shi, X. Zhao, Model predictive control of H5 inverter for transformerless PV systems with maximum power point tracking and leakage current reduction, in: Proceedings: IECON 2018—44th Annual Conference of the IEEE Industrial Electronics Society, 2018, pp. 1860–1865, https://doi.org/10.1109/IECON.2018.8591386.

[21] T.K.S. Freddy, N.A. Rahim, W.P. Hew, H.S. Che, Comparison and analysis of single-phase transformerless grid-connected PV inverters, IEEE Trans. Power Electron. 29 (2014) 5358–5369, https://doi.org/10.1109/TPEL.2013.2294953.

[22] H. Xiao, S. Xie, Y. Chen, R. Huang, An optimized transformerless photovoltaic grid-connected inverter, IEEE Trans. Ind. Electron. 58 (2011) 1887–1895, https://doi.org/10.1109/TIE.2010.2054056.

[23] W. Yu, J.S. Lai, H. Qian, C. Hutchens, High-efficiency MOSFET inverter with H6-type configuration for photovoltaic nonisolated AC-module applications, IEEE Trans. Power Electron. 26 (2011) 1253–1260, https://doi.org/10.1109/TPEL.2010.2071402.

[24] B. Ji, J. Wang, J. Zhao, High-efficiency single-phase transformerless PV H6 inverter with hybrid modulation method, IEEE Trans. Ind. Electron. 60 (2013) 2104–2115, https://doi.org/10.1109/TIE.2012.2225391.

[25] W. Cui, B. Yang, Y. Zhao, W. Li, X. He, A novel single-phase transformerless grid-connected inverter, in: IECON 2011—37th Annual Conference of the IEEE Industrial Electronics Society, 2011, pp. 1126–1130, https://doi.org/10.1109/IECON.2011.6119466.

[26] R. Gonzalez, J. Lopez, P. Sanchis, L. Marroyo, Transformerless inverter for single-phase photovoltaic systems, IEEE Trans. Power Electron. 22 (2007) 693–697, https://doi.org/10.1109/TPEL.2007.892120.

[27] T. Kerekes, R. Teodorescu, P. Rodríguez, G. Vázquez, E. Aldabas, A new high-efficiency single-phase transformerless PV inverter topology, IEEE Trans. Ind. Electron. 58 (2011) 184–191, https://doi.org/10.1109/TIE.2009.2024092.

[28] W. Li, Y. Gu, H. Luo, W. Cui, X. He, C. Xia, Topology review and derivation methodology of single-phase Transformerless photovoltaic inverters for leakage current suppression, IEEE Trans. Ind. Electron. 62 (2015) 4537–4551, https://doi.org/10.1109/TIE.2015.2399278.

CHAPTER 4

Packed U-Cell topology: Structure, control, and challenges

Mohamed Trabelsi[a], Hamza Makhamreh[b], Osman Kukrer[b], and Hani Vahedi[c]

[a]*Electronic and Communications Engineering, Kuwait College of Science and Technology, Kuwait*
[b]*Department of Electrical and Electronics Engineering, Eastern Mediterranean University, Gazimagusa, Turkey* [c]*Ossiaco Inc., Montreal, QC, Canada*

Abbreviations

ANPC	active neutral point clamped
APOD	alternative phase opposite disposition
CHB	cascaded H-bridge
CMPC	classical MPC-controllers
DC	direct current
DVR	dynamic voltage restorer
FC	flying capacitors
FCS-MPC	finite control set model predictive control
FFT	fast Fourier transform
HB	H-bridge
IPD	in-phase disposition
LMPC	Lyapunov-based model predictive control
LSPWM	level-shifted PWM
MLI	multilevel inverter
MPC	model predictive control
NPC	neutral point clamped
POD	phase opposite disposition
PSPWM	phase-shifted PWM
PUC	packed U-cells
PUC5	5-level packed U-cells
PUC7	7-level Packed U-cells
PWM	pulse width modulation
RES	renewable energy source
SDCS-MLI	single-DC-source multilevel inverter
SMC	sliding mode control
THD	total harmonic distortion
UPS	uninterruptible power supply
V2G	vehicle-to-grid

4.1 Introduction

Based on the recent energy statistics, the world's electric energy consumption is increasing continuously, which requires more power generation especially from renewable energy sources (RESs). RESs play an important role in generating power in the form of green energy and with low environmental impacts. However, their raw output is not usable by consumers and needs to be boosted and converted into a smooth AC waveform to deliver the desired power to the grid with low harmonics. Moreover, the industries demand higher power equipment, which are more than megawatt level, such as high-power AC drives, which are usually connected to medium-voltage networks. Thus, a new technology of inverters, called MLIs, has been introduced that utilizes a combination of switches and DC sources to produce various voltage levels, which is being used in medium-voltage high-power applications. The MLI structure allows to generate smoother output waveforms by producing different voltage levels while operating at a lower switching frequency, which leads to less power losses in the power inverter and reduction in the output filter size. Different types of MLIs have been proposed in the literature, which are mainly classified into single and multiple DC source MLIs.

SDCS-MLIs are desired where the auxiliary capacitors are controlled through switching states without adding extra linear/nonlinear regulators and complexity to the system. Such a topology can be installed in all power system applications where the two-level ones are already operating. Therefore, the input DC side and output AC side do not require to be modified. Moreover, the controller remains the same since only a single DC link should be regulated and the error signal is used in the reference current generation. However, the modulation block should be replaced by a multilevel switching technique with integrated voltage balancing using redundant switching states.

The PUC inverter has been recently considered as an interesting single-DC-source multilevel inverter (SDCS-MLI) topology due to its multiple features compared to the other MLI topologies [1,2]. The main benefits of this topology are the flexibility in expanding to higher output levels while utilizing an SDCS, enhancement of the filter bandwidth taking advantage of the switching states redundancy, high reliability, and cost reduction due to the utilization of a reduced number of active components, and high ride-through capability assured by the topology storage capacitors. Currently, the five-level version of PUC is being commercialized in the industry. Modeling, control challenges, and applications for the PUC converter are discussed in this chapter.

4.2 Packed U-cell topology

Despite the abovementioned merits of the PUC converter, the topology suffers from: (1) complexity of the controller design (multi-objective control problem) [3] and (2) hybrid nature, where the control inputs are included in the system matrix [4–7].

These drawbacks make the application of traditional linear and nonlinear control techniques difficult. For instance, in Ref. [8], the authors have used a multicarrier level-shifted pulse width modulator integrated to a voltage regulator in order to balance the capacitor voltage. Making use of the switching states redundancy, the voltage regulator does not imply a direct relation between the controlled variables and the controller parameter(s). Recently, several research works have been published on the application of FCS-MPC to a PUC7 grid-connected inverter [9–11] and a dual output rectifier [12]. In both applications, the cost function is defined by a weighted summation of the system state errors. Thus, the cost function design is omitted from the design procedure and the fine-tuning of the weighting factor is vital to ensure a good performance and system stability. However, a retuning of the weighting factors would be needed for a different system's nominal values (parameters sensitivity) and operating conditions (the weighting factors could be only locally optimized). In Ref. [5], a Lyapunov-based MPC controller is proposed instead. Though the cost function includes errors in three state variables, the proposed controller is characterized by its easy implementation (no gains tuning requirement even under different operation conditions and parameters variation) and lower sensors count (the controller predicts the load currents based on the mathematical model of the PUC7 rectifier). The proposed controller is considered as an improvement of the conventional MPC, where the cost function is derived from a stability point of view. Using the grid current measurement, the authors in Refs. [13,14] proposed a switched observer-based reduced-sensor control of a grid-connected PUC7 inverter.

4.2.1 Mathematical modeling

The PUC inverter is a multilevel topology where each cell consists of one capacitor and two switching devices. Considering n cells, the inverter will consist of $2n$ switches (the two switches of the same cell must be controlled in a complementary way, which will give 2^n combinations with redundant states) and $n - 2$ capacitors. For instance, a three-cell PUC inverter consists of six switching devices, 1DC-Source, and a flying capacitor (reduced number of components for the same number of output voltage levels compared to other MLI topologies) (Fig. 4.1)

The states of the power switches are given as.

$$s_i = \begin{cases} 1, & \text{if } S_i \text{ is ON} \\ 0, & \text{if } S_i \text{ is OFF} \end{cases} \tag{4.1}$$

while the pair of the switching functions (S_a, S_b) is defined as.

$$\begin{aligned} s_a &= s_1 - s_2 \\ s_b &= s_2 - s_3 \end{aligned} \tag{4.2}$$

Table 4.1 Shows the switching patterns for the three-cell PUC inverter.

If the capacitor voltage V_2 is controlled at 1/3 of V_{dc}, then a seven-level (PUC7) output voltage waveform V_{out} would be generated ($\pm 3V_{dc}, \pm 2V_{dc}, \pm V_{dc}, 0$). However, in order to achieve five identical voltage levels at the output (PUC5

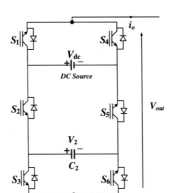

FIG. 4.1

The three-cell PUC inverter configuration.

Table 4.1 Switching patterns of the three-cell PUC inverter.

State (/)	V_{out}	S_1	S_2	S_3
1	V_{dc}	1	0	0
2	$V_{dc}-V_2$	1	0	1
3	V_2	1	1	0
4	0	1	1	1
5	0	0	0	0
6	$-V_2$	0	0	1
7	$V_2 - V_{dc}$	0	1	0
8	$-V_{dc}$	0	1	1

configuration), the voltage at the terminals of the capacitor C_2 must be regulated at $V_2^* = V_{dc}/2$, where V_2^* is the reference voltage for the capacitor C_2. It is worth noting that the PUC5 offers more flexibility in synthetizing the output voltage levels (easy capacitor voltage balancing) compared to its original PUC7 topology due to the redundant switching states (Table 4.1), which leads to an optimized output performance.

Using Kirchhoff's laws, relations between the capacitor voltage $V_2(t)$, the DC-link voltage $V_{dc}(t)$, the output current $i_o(t)$, and the switching states s_i are expressed by Eqs. (4.3)–(4.4):

$$C_2 \frac{dV_2(t)}{dt} = (s_3 - s_2)i_o(t) \qquad (4.3)$$

$$V_{out}(t) = (s_1 - s_2)V_{dc}(t) + (s_2 - s_3)V_2(t) \qquad (4.4)$$

4.2.2 Control challenges

Feedback-based control of the PUC topology requires finding control laws to determine the switching control signals for the switching devices. The control laws should ensure the tracking of the converter output current and the voltage of the auxiliary capacitor in inverter mode/the input current and the capacitor voltages in rectifier mode of their desired values. It has been shown in Section 4.2.1 that the mathematical model of the PUC7 topology makes use of two switching control variables (s_a, s_b) correlated with the controlled variables. Hence, the system model is inherently nonlinear and discrete, meaning that a control variable can take only two values, either zero or one. These control variables are to be selected to set the voltage levels of the converter AC side, thereby controlling the AC side current. Also, one of these variables controls the capacitor current (in inverter mode), which should be controlled to have zero average value in the steady state, since the capacitor voltage is DC in the steady state. A major difficulty with such a system is that state-space averaging cannot be directly applied to the system model to obtain a continuous-time approximation for the control variables, from which the control signals can be generated by a PWM process. This renders the application of classical control techniques to this topology very difficult. Another setback, which is related to the averaging issue, is the almost impossibility of linearizing the system around an operating point. In other words, the control signals corresponding to a specified steady state of the system cannot be easily calculated. Without state-space averaging and linearization, calculation of such an operating point for the system is an intractable problem.

4.3 Control techniques

As mentioned above, the main challenge in the PUC control is to best track the output current reference while maintaining the capacitor voltage at its nominal value. It is also worth noting that the controlled variables are interrelated and any change in one of them may affect the others. This means that if the PUC smoothly controls one of the above variables, it may cause degradation in the control of others.

Therefore, one should be very careful in the choice of the controller type and its parameters. Different types of controllers were proposed in the literature [1] (Fig. 4.2). The model predictive control (MPC) strategy has been considered as the most appropriate control approach for the PUC topology by researchers till recently. The MPC strategy, as applied to the PUC topology, selects the switching control variables to minimize a properly designed cost function, which is calculated using predicted values of the controlled variables at the end of the sampling interval. The predictions are based on the mathematical model of the converter. The cost function is evaluated for all feasible pairs of the control variables, and the pair which leads to the minimum of the cost function is selected. Other approaches have been proposed in the literature based on PWM (pulse width modulation) representation of the switching control signals using multiple carrier signals. The drawbacks of this

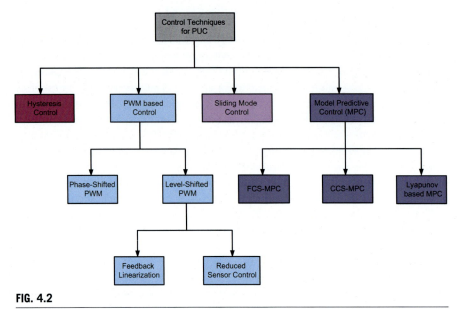

FIG. 4.2

Control techniques used in PUC topologies.

approach have been pointed out in the paragraph above. Hence, the PWM control approach needs further elaboration and mathematical verification.

Although the MPC approach has been quite successful in achieving the control objectives, several challenging issues remain to be solved. Selection of the gains in the cost function, which determine the relative weights of the state variables in the cost, cannot be based on well-defined criteria in terms of the performance of the system. Another problem concerns the switching frequency of the converter. As a consequence of the MPC control strategy, the switching frequency cannot be readily predicted or controlled and is in general variable during an output cycle. In MPC, it is possible to apply some control on the switching frequency by including a term in the cost function that represents a limitation on the switching frequency. Such additional terms in the cost function usually complicate the implementation of the MPC strategy. In the following sections, two approaches published in the literature are described that attempt to solve these problems. These approaches themselves are based on MPC. However, the cost function is derived, in one case, from Lyapunov stability, and from sliding mode control principles in the other case.

4.3.1 Finite set model predictive control

Let the capacitor voltage $V_2(t)$ and the grid current $i_g(t)$ be the two state variables of the grid-connected PUC7 inverter shown in Fig. 4.3. The main idea of the MPC scheme implemented in Ref. [9] is the prediction of the state variables (the grid

4.3 Control techniques

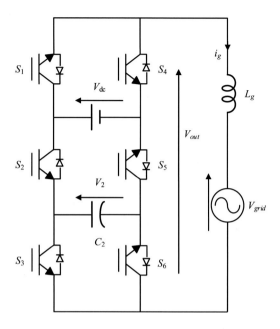

FIG. 4.3

The grid-connected PUC7 inverter.

current i_g^{k+1} and the capacitor voltage V_2^{k+1}) for each possible switching state by means of the discrete-time equations of the system state variables.

Since the state trajectories could be considered as rectilinear for a small sampling time, the authors made use of a simplified model. With this hypothesis, the state variables are approximated for each sampling time T_s, by.

$$x^{k+1} = x^k + \dot{x}(t).T_s \tag{4.5}$$

Then the prediction of the state variables at the $(k+1)$ sample in terms of the measurements at the actual (k) sample can be expressed by Eqs. (4.6)–(4.7):

$$V_2^{k+1} = V_2^k + \frac{T_s}{C_2}(s_3 - s_2)i_g^k \tag{4.6}$$

$$i_g^{k+1} = i_g^k + \frac{T_s}{L_g}\left[(s_1 - s_2)V_{dc}^k + (s_2 - s_3)V_2^k - V_{grid}^k\right] \tag{4.7}$$

Usually, after the prediction of the state variables, the MPC techniques select the switching state giving the minimal error between the reference and the measurement. Therefore, these techniques are not taking into account the dissimilarity of the variation ranges (hundreds of Volts for voltages and a few Amperes for current), which leads to less tracking capabilities. Thus, the authors in Ref. [9] proposed a normalization of the state variables by calculating the maximal variations ΔV_{2max}, and Δi_{gmax} of the state variables as.

$$\Delta V_{2max} = \frac{2i_g}{C_2}T_s$$
$$\Delta i_{gmax} = \frac{2V_{dc}}{L_g}T_s \tag{4.8}$$

The proposed cost function has the objective of minimizing the error between the predicted grid current i_g^{k+1}, the capacitor voltage V_2^{k+1}, and their references. This control objective is represented as follows:

$$g = \lambda \left| \frac{V_2^* - V_2(k+1)}{\Delta V_{2max}} \right| + \left| \frac{i_g^* - i_g(k+1)}{\Delta i_{gmax}} \right| \tag{4.9}$$

where λ is a weighting factor (which can be adjusted according to the desired performance), V_2^* is the desired voltage across capacitor C_2 ($V_2^* = V_{dc}/3$), and i_g^* is the reference grid current. It should be mentioned that this weighting factor has a crucial effect on the quality of the controlled variables.

Then an experimental study was conducted on the grid-connected PUC7 inverter using the parameters given in Table 4.2. Figs. 4.4 and 4.5 show the experimental results for the tuned scenario ($\lambda = 2 \times 10^{-1}$). Referring to the bottom waveforms of Fig. 4.4, the grid current is perfectly tracking its sinusoidal reference. The waveforms of the upper part of Fig. 4.4 show the generated seven-level output voltage with the grid voltage. The lower part of Fig. 4.5 shows that the capacitor voltage V_2 is oscillating around its reference value ($V_{dc}/3$). This proves the voltage balancing capability of the proposed controller.

4.3.2 Multicarrier pulse width modulation

The PWM technique for multilevel inverters uses the same concept as conventional two-level ones but includes slight modifications [15]. The n-level multilevel inverter needs $n - 1$ firing pulses in each phase without considering the complementary ones. For instance, a full-bridge single-phase inverter with four switches needs only 1 principle pulse to run all switches to generate a two-level voltage waveform at the output. However, the same configuration can be run by two separate pulses that result in

Table 4.2 Grid-connected PUC7 system parameters used for the MPC implementation.

Parameters	Symbol	Value
Dc voltage source	V_{dc}	150 V
Capacitor	C_2	100 µF
Grid voltage (peak) and frequency	V_g, f	120 V, 50 Hz
Grid inductance	L_g	10 mH
Sampling frequency	F_s	10 kHz

4.3 Control techniques

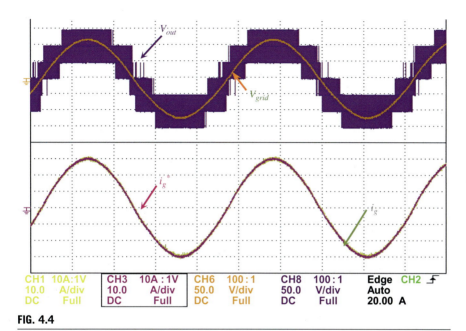

FIG. 4.4

Experimental results for tuned MPC, Upper: output and grid voltage; Lower: grid current.

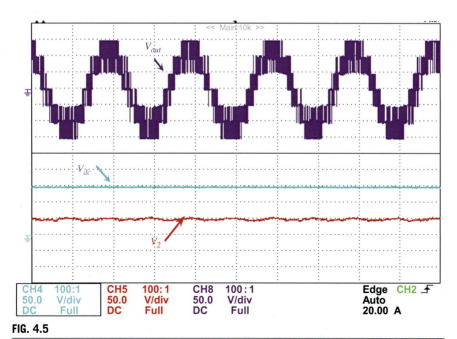

FIG. 4.5

Experimental results for tuned MPC, Upper: output voltage; Lower: capacitor C2 voltage.

forming a three-level voltage at the output. Consequently, a five-level inverter needs four firing pulses per phase. The PWM technique has been modified and developed to generate separate pulses from one sinusoidal wave. These developed PWM methods are known as level-shifted and phase-shifted techniques which are described as the following [16].

4.3.2.1 Level-shifted PWM
As explained above, for an *n*-level multilevel inverter, $(n - 1)$ carrier waves would be compared to a sinusoidal wave. In a two-level PWM, there is only a single carrier with an amplitude range of $-V$ to $+V$, while in a level-shifted PWM (LSPWM), the carrier waves are divided equally within that range. Therefore, the amplitudes of $n - 1$ carriers are $(1/(n-1))^{th}$ times of the carrier in a two-level PWM. These carrier waves are vertically disposed such that the bands they occupy are adjacent. There are various methods in moving the carriers vertically:

- In-phase disposition (IPD): all carriers are in phase;
- Alternative phase opposite disposition (APOD): all carriers are alternatively in opposite disposition;
- Phase opposite disposition (POD): all carriers above the zero reference are in phase but in opposition with those below the zero reference.

Fig. 4.6 gives an example of these three types of carrier disposition in level-shifted PWM for a five-level PWM with four carrier waves.

For instance, consider a five L-CHB that has two HBs in each leg and each HB needs two pulses. V_{cr1} and V_{cr2} are for the first HB, and similarly, V_{cr3} and V_{cr4} are for the second HB. Fig. 4.6A–C, respectively, show the positions of carriers in IPD, APOD, and POD that have been defined above. The switching pulses will be generated by comparing the carriers with the modulation wave similarly to the main PWM method.

4.3.2.2 Phase-shifted PWM
In the phase-shifted PWM technique, $n - 1$ triangular waves would have the same frequency and amplitude as the carrier in a two-level PWM, but there is a phase shift between adjacent carrier waves which is.

$$\phi_{cr} = \frac{360°}{n-1} \tag{4.10}$$

The carrier waves for a five-level inverter are drawn in Fig. 4.7. So these carriers should have a 90° phase shift, consecutively. These carriers are shown in the figure by V_{cr1}, V_{cr2}, V_{cr3}, and V_{cr4}. As the phase difference is 90°, so a 180° phase shift is seen between V_{cr1} and V_{cr3} as well as between V_{cr2} and V_{cr4}.

Except the procedure of carrier waves, the other steps of generating pulses are like the level-shifted PWM. It should be mentioned that this technique does not have good results for NPC multilevel inverters and is just suitable for CHBs. However, generally, the phase shift PWM has better performance in eliminating the switching harmonics.

4.3 Control techniques

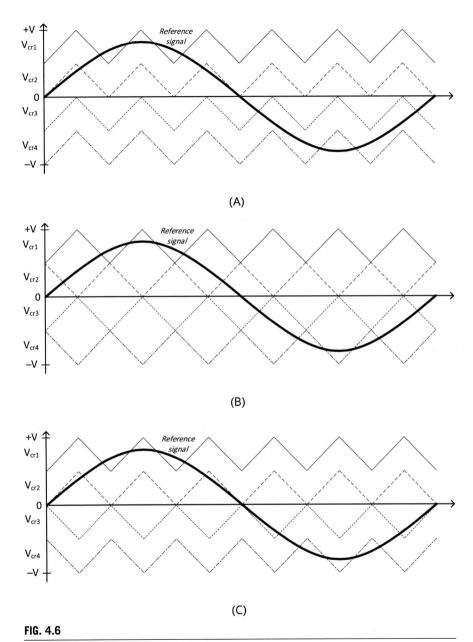

FIG. 4.6

Level Shifted PWM Carriers for 5L Inverter: (A) IPD, (B) APOD, (C) POD.

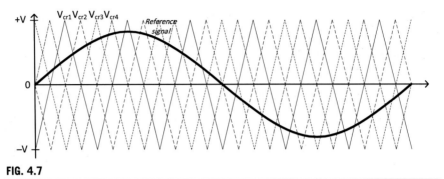

FIG. 4.7

Phase Shifted PWM Carriers for 5L Inverters.

It should be noted that a multicarrier PWM helps in multiplying the switching harmonics by the number of levels. For instance, if the switching frequency is set to m kHz, the first dominant harmonic component will be observed at the $[(n-1) \times m]$th order which reduces the output filter size significantly.

4.3.2.3 Experimental results and discussion

Some tests have been conducted on a five-level CHB inverter with RL load (Fig. 4.8). In these tests, all four carriers' frequencies have been set at 1.2 kHz. The results have been captured as shown in Fig. 4.9. They contain each H-bridge output voltage as well as the main output voltage and current supplying the RL load.

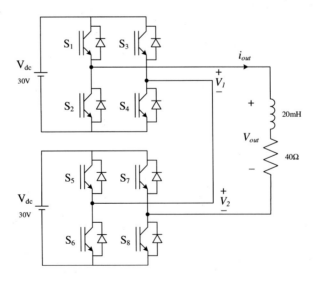

FIG. 4.8

A 5-Level CHB inverter with equal DC sources.

4.3 Control techniques

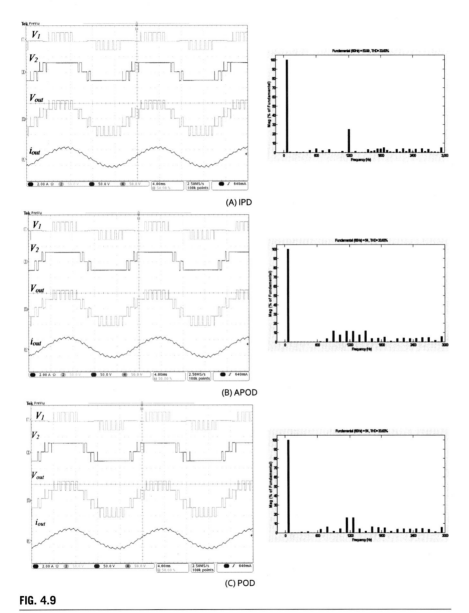

FIG. 4.9

Results of applying Level-shift PWM on 5-level CHB.

A close look at the voltage waveforms makes it clear that the voltage pulses are a bit different. However, considering the total harmonic distortion (THD) of the output voltage/current waveform, all results of different level-shifted PWM techniques are almost the same. It should be mentioned that the FFT analysis of the output voltage waveform revealed the first dominant harmonic at 1200 Hz, which is the carrier frequency, but the actual switching frequency can be measured from Fig. 4.9 at 300 Hz by counting the on and off pulses in waveforms, which is 1/4th of the carrier frequency. It can be observed that the 1200 Hz harmonic component and its subharmonics are different in those three types of level-shift PWM technique.

Fig. 4.10 shows the result of the same inverter running by phase-shifted PWM. In this case, the carriers' frequencies have been set at 300 Hz to get the same results as level-shifted PWM. Therefore, the harmonic spectrum showed the first dominant harmonic at four times higher than 300 Hz, which is measured at 1200 Hz. By counting the number of on and off pulses, it is obvious that the switches are operating at 300 Hz.

One main difference is that the two H-bridge output cells are identical in phase shift PWM while they are different in level-shift one. Consequently, the power losses would be equal among switches applying phase shift PWM. However, there are some methods reported to modify the level-shift PWM in order to get the equal voltage pulses from different cells of the CHB inverter [17]. Considering the FFT analysis, it can be seen that the first dominant harmonic is placed around the 1200 Hz and the harmonic spectrum looks like the APOD level-shift PWM one.

4.3.3 Lyapunov-based model predictive control

Model predictive control (MPC) offers a simple and effective control method for renewable energy-related applications with remarkable ease of implementation [9,18,19]. The presented Lyapunov-based model predictive controller has a cost function that is derived intuitively based on the Lyapunov control theory [4]. When compared to classical MPC controllers (CMPC), the proposed controller predicts the states based on the mathematical model of the system and evaluates a cost function (a combination of the system's errors) for all the control input pairs for a minimum error value. The MPC algorithm minimizes a cost function, where the state variables might be not homogeneous (Ampere and Volt) with very different variation ranges

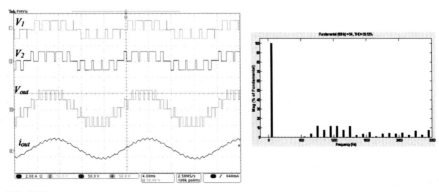

FIG. 4.10

Results of applying Phase-shift PWM on 5-level CHB.

4.3 Control techniques

(hundreds of Volts for voltages and a few Amperes for current). Thus, usually, a normalization of the state variables and inclusion of a weighting factor are needed to achieve the desired performance. Yet LMPC differs in the form of the cost function used in the controller. Thus, the proposed cost function eliminates the need for optimization and tuning of the weighting factors associated with the states' error. Due to the usage of a Lyapunov-based cost function, the stability of the controlled system is expected to be guaranteed.

Referring to the mathematical modeling presented in Section 4.2.1, the capacitor dynamic is expressed as:

$$i_c = C\frac{dv_c}{dt} = -s_b i_g \tag{4.11}$$

while the grid current dynamic is given by.

$$\frac{di_g}{dt} = -\frac{r_g}{L_g}i_g + \frac{1}{L_g}(v_i - v_g) \tag{4.12}$$

Let us define $x_1 = i_g - i_g^*$ and $x_2 = v_c - v_c^*$ as the grid current and the capacitor voltage errors, where i_g^* and v_c^* represent the grid current and the capacitor voltage references, respectively. Additionally, v_{dc} is the DC supply voltage, which is three times the value of the capacitor voltage reference.

From the above equations, one can obtain the current error derivative as.

$$\dot{x}_1 = \frac{1}{L_g}\{s_a v_{dc} + s_b(x_2 + v_c^*) - r_g x_1 - v_i^*\} \tag{4.13}$$

and the capacitor voltage error as.

$$\dot{x}_2 = -\frac{1}{C}s_b(x_1 + i_g^*) \tag{4.14}$$

where v_i^* is the reference output voltage of the inverter (assuming that the reference current is injected into the grid) given as.

$$v_i^* = v_g + r_g i_g^* + L_g \frac{di_g^*}{dt} \tag{4.15}$$

The basic idea of this control method is the introduction of the Lyapunov cost function instead of the weighted error cost function used in the classical MPC control.

The Lyapunov theory states that if an energy cost function (called the candidate Lyapunov function) is appropriately chosen to control the system, then the system stability is guaranteed. The control input is selected so as to make the derivative of the candidate function negative. The candidate function has to: (1) be positive definite, (2) have a derivative which is negative definite and, (3) go to infinity as the states go to infinity. In the present model, the candidate Lyapunov function is defined as.

$$V(x) = \frac{1}{2}a_1 x_1^2 + \frac{1}{2}a_2 x_2^2 \tag{4.16}$$

where the factor of 1/2 is chosen arbitrarily to simplify the derivation of the following steps. The gains α_1 and α_2 are real positive numbers. To apply the Lyapunov control theory to the PUC7 model, we need to select the control input pair, which makes the derivative of the Lyapunov control function in Eq. (4.16) negative. Taking the derivative of Eq. (4.16) and substituting Eqs. (4.13) and (4.14) in the derivative equation gives:

$$\dot{V}(\mathbf{x}) = \frac{1}{L_g} s_b x_1 x_2 \left(\alpha_1 - \frac{L_g}{C} \alpha_2 \right) - \frac{\alpha_2}{C} x_2 s_b i_g^* + \frac{\alpha_1}{L_g} x_1 \left(s_a v_{dc} + s_b v_c^* - r_g x_1 - v_i^* \right) \quad (4.17)$$

The gains can be selected to eliminate the $x_1 x_2$ term in the above equation, which can be written as.

$$\dot{V}(\mathbf{x}) = \frac{\alpha_1}{L_g} \left\{ x_1 \left(s_a v_{dc} + s_b v_c^* - r_g x_1 - v_i^* \right) - x_2 s_b i_g^* \right\} \quad (4.18)$$

where α_1 can be any real positive number.

The derivative of the Lyapunov function is evaluated at the instant $(k+1)$ and referred to as the cost function given by:

$$\dot{V}_\mathbf{x}^{(l)}(k+1) = \frac{\alpha_1}{L_g} \{ x_1(k+1) \{ s_a^{(l)}(k) v_{dc} + s_b^{(l)}(k) v_c^* - r_g x_1(k+1) - v_i^*(k+1) \} \\ - x_2(k+1) s_b^{(l)}(k) i_g^*(k+1) \} \quad (4.19)$$

where l refers to the level index (or the state) in Table 4.1. After measuring the grid current, the grid voltage, and the capacitor voltage, the cost function is calculated for all the possible input pairs ($l = 1...7$), and the controller targets the minimum value of the cost function given by Eq. (4.19). Basically, the controller works as a conventional MPC but with a cost function that does not involve gain tuning parameters (the value of α_1 in Eq. (4.19) is set to 1, or any positive value). Based on the experimental parameters given in Table 4.3, the steady-state performance of the proposed controller is shown in the first two cycles, and the dynamic performance is shown in the last two cycles due to a step change in the current reference (Fig. 4.11).

Table 4.3 Experimental parameters of the implementation of the Lyapunov-based MPC.

Parameters	Symbol	Value
Dc voltage source	V_{dc}	210 V
Capacitor	C	1.5 mF
Grid voltage (RMS) and frequency	V_g, f_g	120 V, 60 Hz
Grid inductance and resistance	L_g, r_g	5 mH, 0.7 Ω
Sampling time	T_s	25 µs

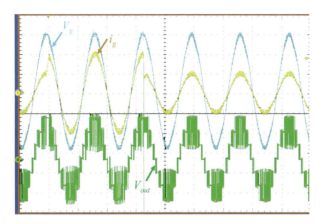

FIG. 4.11

Steady state and dynamic response of the proposed Lyapunov-based MPC.

4.3.4 Sliding mode control

The classical ways of designing the sliding mode controller are not suitable for the PUC topology due to its bilinear structure (hybrid nature), where the control inputs s_1 and s_2 are included in the control matrix, as discussed in Section 4.2.2. Usually, when a zero state error is targeted, the designer defines the switching function as a combination of the error states (grid current error x_1 and capacitor voltage error x_2 in here). Afterward, the control action is selected in order to drive the switching function to zero. A possible sliding function is given by Refs. [6,7]:

$$\sigma(\mathbf{x}) = \alpha x_1 + x_2 \quad (4.20)$$

Here the parameter α is a weighting gain (in ohms) where $\mathbf{x}$ represents the error vector. The sliding mode occurs on the line $\sigma = 0$. On the $x_1 x_2$- plane, the state moves from any initial point on the plane to the sliding line ($\sigma = 0$); this phase is called the reaching mode, where the state trajectory reaches the sliding line. Once it hits the sliding line, the state "slides" on the sliding line till it reaches the steady state. When the system reaches the steady state, this implies that a zero error is maintained. For the stability of the reaching mode, the condition $\sigma \dot{\sigma} < 0$ should be satisfied.

The controller design requires to have a control law in terms of the switching function and system parameters in order to achieve the reaching condition.

$$(s_1, s_2) = f(\sigma, \dot{\sigma}, r_g, L_g, C) \quad (4.21)$$

The challenge here is to find the function f in the above equation. Furthermore, in conventional SMC design problems, the system dynamical equations should be in a form such that x_2 is directly linked to the time derivatives of x_1 ($\dot{x}_1, \ddot{x}_1$, etc.) so that the sliding mode can be designed to be stable by properly choosing the parameter α. In the PUC inverter case, x_2 cannot be directly linked to $\dot{x}_1$ and vice versa. Thus, one of the main contributions of this control method is to propose a new "multi-sliding surface" concept to overcome the complexity of finding the switching function for hybrid systems (PUC topology). The sliding function in Eq. (4.20) is substituted by two separate switching functions designed, for each error state, to overcome the problem of the hybrid nature of the PUC topology. The switching functions are chosen as

$$\begin{aligned}\sigma_1 &= x_1 \\ \sigma_2 &= x_2,\end{aligned} \quad (4.22)$$

which have reaching conditions given by.

$$\begin{aligned}\sigma_1 \dot{\sigma}_1 &< 0 \\ \sigma_2 \dot{\sigma}_2 &< 0\end{aligned} \quad (4.23)$$

The negativity of the product of a switching function with its derivative ensures that the state trajectory is directed to the sliding line ($\sigma_1 = x_1$ and $\sigma_2 = x_2$). The states are eventually driven to the steady-state point (0,0).

Fig. 4.12 shows the SMC algorithm. The reaching conditions in Eq. (4.23) can be written as

$$\begin{aligned}w_1^{(l)}(k) &= \frac{x_1(k)}{L_g} \left\{ s_1^{(l)}(k) v_{dc} + s_2^{(l)}(k) v_c^* + s_2^{(l)}(k) x_2(k) - v_i^*(k) \right\} \\ w_2^{(l)}(k) &= \frac{-x_2(k)}{C} s_2^{(l)}(k) i_g(k)\end{aligned} \quad (4.24)$$

where l represents the index in Table 4.1.

The error in the capacitor voltage is bounded within a small hysteresis band, the width of which is denoted by h as presented in Fig. 4.12. The experimental results of the proposed algorithm are obtained based on Table 4.4, where the effect of the hysteresis width choice can be seen from the comparison of Fig. 4.13, which depicts the steady-state performance when $h = 0V$, and Fig. 4.14 when $h = 1V$. It is clearly shown that the selection mechanism greatly reduces the average switching frequency. Though this can also be done when using FCS-MPC, yet SMC has the advantage of directly coupling the capacitor voltage error and the hysteresis width value. In other words, the choice of λ in Eq. (4.9) does not give an idea about the expected capacitor voltage error, where the SMC controller can directly relate both values.

4.3.5 Reduced sensor control

The use of a DC capacitor in the PUC5 topology makes the voltage control mandatory as for any other MLI topology. Motivated by the importance of the correct knowledge of the capacitors' voltages in the control design (Table 4.5), recent research works were focused on the accurate estimation of the capacitors' voltages for different PUC5 configurations [8,20–23]. The proposed sensorless voltage technique reduces the complexity of the control system, which makes the PUC5 inverter appealing for industrial applications.

4.3 Control techniques

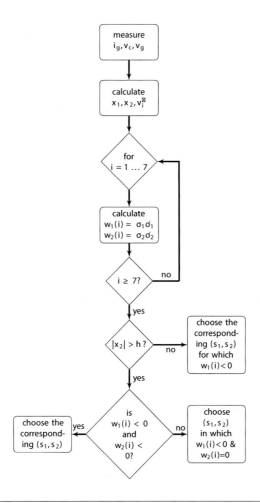

FIG. 4.12

Flowchart of FCS-SMC algorithm.

Table 4.4 Grid-connected PUC7 system parameters used for the MPC implementation.

Parameters	Symbol	Value
Dc voltage source	V_{dc}	150 V
Capacitor	C_2	100 µF
Grid voltage (peak) and frequency	V_g, f	100 V, 50 Hz
Grid inductance	L_g	10 mH
Sampling time	T_s	25 µs

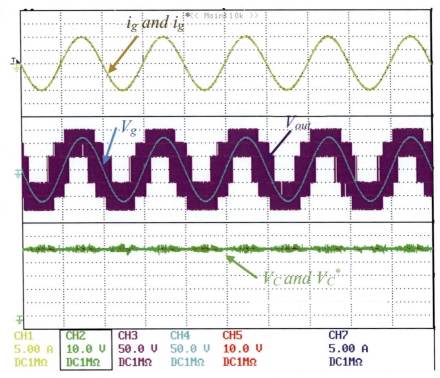

FIG. 4.13

Experimental performance of SMC algorithms without hysteresis ($h = 0$).

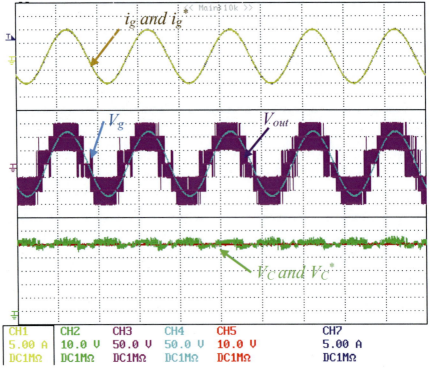

FIG. 4.14

Experimental results of FCS-SMC during steady state ($h = 1V$).

4.3 Control techniques

Table 4.5 States of charge of the PUC capacitor.

S_1	S_2	S_3	C_2
1	0	0	Unchanged
1	0	1	Charging
1	1	0	Discharging
1	1	1	Unchanged
0	0	0	Unchanged
0	0	1	Discharging
0	1	0	Charging
0	1	1	Unchanged

By regulating the DC capacitor voltage at half of the DC source value, the authors in Ref. [8] proposed a PWM-based sensorless voltage control for a PUC5 inverter (Fig. 4.15). Four level-shifted carriers (Cr_1 to Cr_4) are compared to the modulation signal V_{ref} to produce the five-level PWM scheme. The two configurations shown in Fig. 4.16 are used to balance the capacitor voltage at the desired level. The steady-state relationships between V_1 and V_2 are derived by Eqs. (4.25)–(4.34).

The capacitor current should be used to analyze its voltage balancing principle. The current i_c is equal to i_s during the switching states in which the capacitor is involved. Therefore, the following equations could be written for that current:

$$i_c = i_s \tag{4.25}$$

$$v_L = L \frac{di_s}{dt} \tag{4.26}$$

The capacitor charge balance can be written as.

$$\int_{char.} i_s\, dt + \int_{dischar.} i_s\, dt = 0 \tag{4.27}$$

The capacitor voltage V_2 is assumed to be ripple free.

Assume the charging and discharging times t_p and t_n, respectively, as shown in Fig. 4.17.

During the charging time, Fig. 4.16A and Eq. (4.26) give.

$$i_s = \frac{1}{L}\int_0^t (V_1 - V_2 - v_s)\, dt + i_{so} = \frac{V_1 - V_2}{L}\int_0^t dt - \frac{1}{L}\int_0^t v_s\, dt + i_{so} \tag{4.28}$$

Leading to.

$$i_s = \left(\frac{V_1 - V_2}{L}\right) t + i_{so} - \frac{1}{L}\int_0^t v_s\, dt \tag{4.29}$$

During the discharging time, and assuming the time origin is now at the beginning of this interval, Fig. 4.16B and Eq. (4.26) lead to.

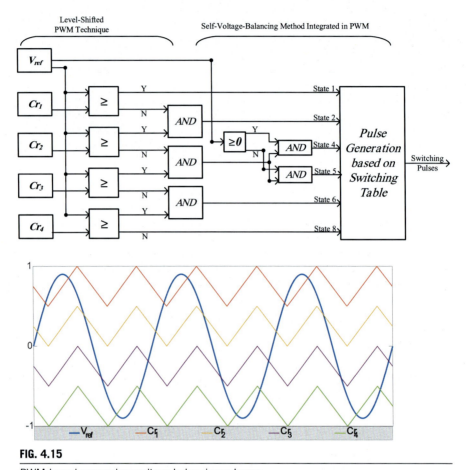

FIG. 4.15

PWM-based sensor-less voltage balancing scheme.

$$i_{s'} = \frac{1}{L}\int_{o'}^{t'}(-V_2 - v_s)\,dt' + i_{so'} = \left(-\frac{V_2}{L}\right)t' + i_{so'} - \frac{1}{L}\int_{o'}^{t'} v_s\,dt' \qquad (4.30)$$

where i_{so} and $i_{so'}$ are the initial currents of charging and discharging intervals, respectively. Applying the charge balance as in Eq. (4.27) yields

$$\int_{o}^{t_p} i_s\,dt + \int_{o}^{t_n} i_{s'}\,dt' = 0 \qquad (4.31)$$

$$\frac{V_1 - V_2}{2L}t_p^2 - \frac{V_2}{2L}t_n^2 + i_{so}t_p + i_{so'}t_n - \frac{1}{L}\left[\int_{o}^{t_p}\left(\int_{o}^{t} v_s\,dx\right)dt + \int_{o}^{t_n}\left(\int_{o'}^{t'} v_s\,dx\right)dt'\right] = 0 \qquad (4.32)$$

4.3 Control techniques

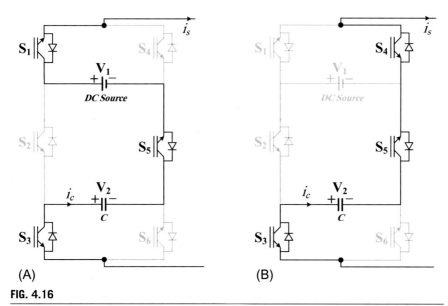

FIG. 4.16

PUC5 capacitor state of charge (A) charging, and (B) discharging.

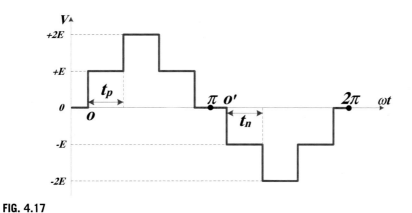

FIG. 4.17

Output 5-level voltage waveform and typical charging/discharging intervals.

The sinusoidal shapes of current i_s and voltage v_s, as well as symmetry in the control processed error (as shown below), imply that

$$i_{so} = -i_{so'} \quad \text{and} \quad \int_o^{t_p}\left(\int_o^t v_s\, dx\right) dt + \int_o^{t_n}\left(\int_{o'}^{t'} v_s\, dx\right) dt' = 0 \tag{4.33}$$

Thus, the charge balance expression simplifies to

$$\frac{V_1-V_2}{2L}t_p^2 - \frac{V_2}{2L}t_n^2 + i_{so}(t_p - t_n) = 0 \tag{4.34}$$

Furthermore, it is obvious that, owing to the half-wave symmetry in the reference signal, to every charging duration t_p corresponds an equal discharging time duration t_n.

Considering the above fact, V_{ref} has a half-wave symmetry as a sine wave; it is then clear that the switching states two and six would have equal intervals. Going back to Eq. (4.34), the following relation is achieved:

$$V_1 = 2V_2 \tag{4.35}$$

One can conclude that, by using the proposed switching technique, the capacitor voltage is regulated around its reference value. Fig. 4.18 shows the experimental result of the sensorless voltage balancing technique applied on the PUC5 inverter. As depicted in that figure, the DC bus is started at 200 V immediately and then the capacitor voltage is tracking to reach the 100 V level. Consequently, the five-level voltage waveform is formed at the output smoothly. Another test has been performed to investigate the performance of the sensorless voltage balancing technique. As shown in Fig. 4.19, a harmonic load (diode rectifier) has been added to the existing load of the PUC5 and the auxiliary capacitor voltage is still balanced with no negative effect.

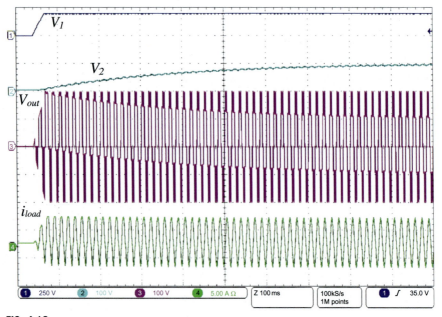

FIG. 4.18

Soft start of the PUC5 inverter with voltage balancing of the auxiliary capacitor.

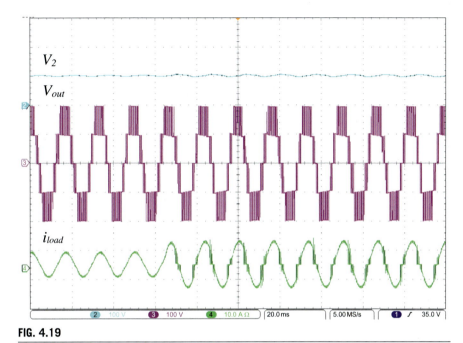

FIG. 4.19

Adding single-phase rectifier (as nonlinear load) paralleled with the RL load to the output of PUC5.

The proposed sensorless voltage controller has been integrated into the switching technique to work as an open-loop system with reliable results. Since the voltage balancing technique has been integrated into the switching pattern through modifying the multicarrier PWM, there is no need for an external controller which increases the order and complexity of the system and reduces the stability and reliability consequently. The low harmonic components in both voltage and current waveforms generated by PUC5, no need for bulky output filters, reliable and good dynamic performance in variable conditions (including change in DC source, load, power amount injected to the grid), no voltage/current sensor requirement in stand-alone mode, low manufacturing costs, and miniaturized package due to using less components are the interesting advantages of the introduced PUC5 topology which have been proved by experimental results.

4.4 Applications

As mentioned in the previous sections, the PUC5 is a modification to the original seven-level configuration with less voltage levels but higher reliability and controllability. Thus, in this section, we are considering the different applications for the PUC5 inverter.

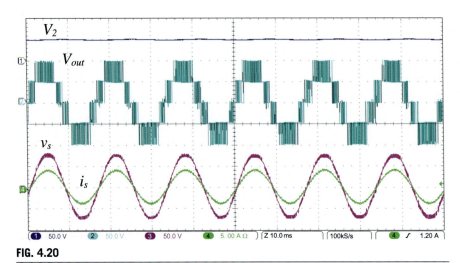

FIG. 4.20

Sensor-less 5-Level PUC grid-connected inverter voltage and current waveforms.

4.4.1 Stand-alone mode

In the stand-alone operation mode (UPS, vehicle-to-home, …), the capacitor voltage (V_2) is automatically regulated without using any voltage sensor and a five-level output voltage waveform is achieved. Fig. 4.19 illustrates the corresponding experimental results, where a nonlinear load (rectifier) is paralleled to the RL load.

4.4.2 Grid-connected mode

The abovementioned sensorless technique was applied to a grid-connected PUC5 inverter and the experimental results are depicted in Fig. 4.20. The presented results show a five-level operation while injecting current into the grid with a unity power factor and regulating the capacitor voltage around its reference value. The presented results make the PUC5 inverter a good candidate for renewable energy and Vehicle-to-Grid (V2G) applications.

4.4.3 PUC5 rectifier

The experimental results shown in Fig. 4.21 present the performance of the PUC5 rectifier. One can notice that the output capacitor voltage is controlled to its reference value. At the source side, the input current is sinusoidal (low harmonic content) and in phase with the source voltage (unity power factor operation), while the five-level input voltage waveform helps in reducing the input filter size. Thus, the PUC5 rectifier could be an interesting candidate for many industrial applications such as traction systems and electric vehicle battery chargers.

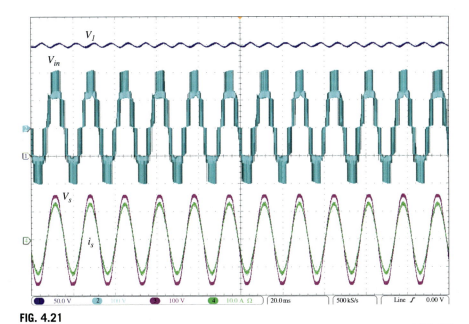

FIG. 4.21

Experimental results of a PUC5 rectifier.

4.4.4 PUC5-based STATCOM

The previously described sensorless technique was also used in the control of a PUC5-based STATCOM. The aim is to inject a controlled amount of reactive power to the grid while regulating capacitor voltages. The results presented in Fig. 4.22 show a 90° phase shift between the grid voltage and the injected current (reactive power injection). Thus, the PUC5 STATCOM could be an interesting alternative to the usually installed bulky capacitors in power/PV systems to inject the required reactive power at any level.

4.4.5 PUC5-based DVR

In power systems, voltage swell/sag, harmonic content, islanding, and long interruptions may have a considerable impact on the overall system power quality. In general, the grid voltage variations (voltage sags/swells) are considered as grid voltage disturbances. Thus, power electronics components such as dynamic voltage restorers (DVRs) are usually used to compensate for these voltage disturbances. The DVR is an active component (inverter) that compensates for the grid voltage disturbances by injecting, at the point of common coupling, a series voltage component synchronized with the grid. Taking advantage of the voltage storage capability of its flying capacitor (voltage ride-through capability), a PUC5-based DVR (Fig. 4.23) was proposed in Ref. [24]. A multi-objective MPC strategy was proposed to control the filter

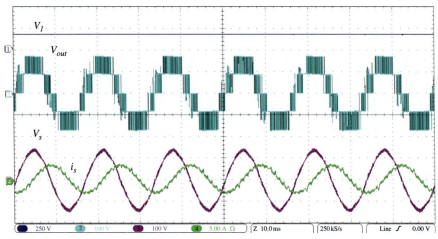

FIG. 4.22

Experimental results of PUC5 STATCOM.

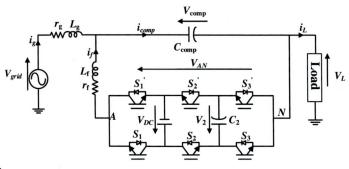

FIG. 4.23

Proposed control synoptic for the PUC5-based DVR.

current, the capacitor voltage, and the compensation voltage. The simulation results presented in Fig. 4.24 depict the grid voltage, load voltage, and compensation voltage during a voltage sag of five cycles. As can be noticed from the figure, before the voltage sag event, the compensation voltage is maintained around zero. When a voltage sag event occurs, the required compensation voltage is generated to keep the desired voltage level constant and the sensitive load unaffected.

4.4.6 PUC5 three-phase inverter

Multilevel three-phase inverters have been mainly finding applications in high-power UPS systems, motor drives, and traction systems. Multilevel three-phase inverters are preferred to conventional two-level inverters due to their improved waveforms quality (lower THD). Thus, a three-phase PUC5 inverter (three and four

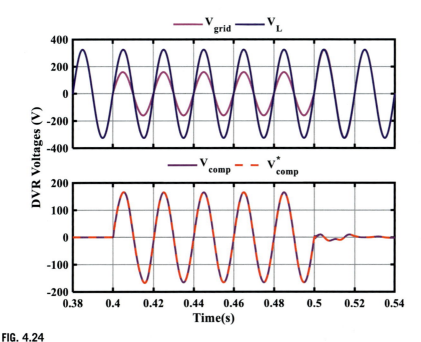

FIG. 4.24

Dynamic performance of the PUC5-based DVR during a grid voltage sag.

wire connections are possible) was developed by connecting three single-phase PUC5 units (Fig. 4.25). Only three isolated DC sources are required, which represents a considerable advantage compared to other three-phase MLI topologies. Fig. 4.26 shows the steady-state waveforms of the employed stand-alone three-phase PUC5 inverter.

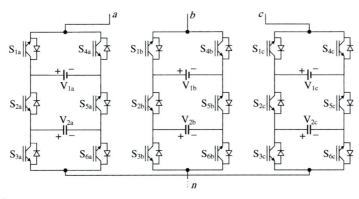

FIG. 4.25

Three-phase PUC5 inverter configuration.

CHAPTER 4 Packed U-Cell topology: Structure, control, and challenges

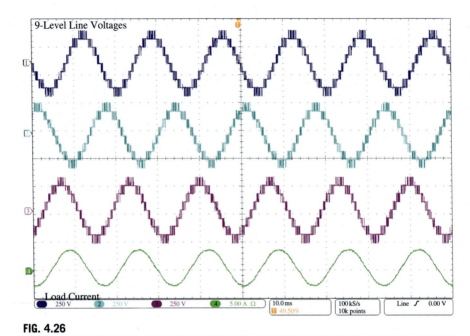

FIG. 4.26

Steady-state waveforms of the three-phase PUC5 inverter.

The PUC5 inverter has been analyzed based on the topology and control design. It has been compared to some mostly used topologies to reveal the promising features of that configuration. The main advantage of the PUC5 inverter is the redundant switching states, which eliminate the need for an external voltage regulator of an auxiliary capacitor. The sensorless voltage balancing technique applicable on the PUC5 inverter has been explained theoretically. It was shown that the auxiliary capacitor could be controlled without using any voltage feedback surprisingly. Moreover, the detailed analysis of single-DC-source and multiple-DC-source topologies along with the associated controller designs indicated the fact that single-DC-source converters are much more interesting for power industries due to the lower manufacturing price and less complicated controllers. It has been demonstrated that the conventional cascaded controller can be simply applied on the PUC5 converter without additional regulators since the auxiliary capacitor voltage is regulated through redundant switching states acceptably. Eventually, the full assessment among FB, CHB, NPC, T3, FC, ANPC, and PUC5 topologies proved that the presented configuration is a potential candidate to compete in the market of power electronics converters and a full range of applications is expected for the PUC5 inverter as shown in Fig. 4.27.

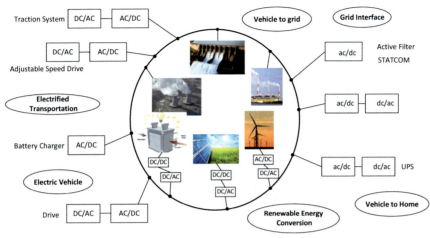

FIG. 4.27

Possible applications of the PUC5 inverter

4.5 Commercialization challenges

The path from an idea, as innovative as it may be, to a successful commercial product is a long and difficult one. PUC5 commercialization is a good example. It can be analyzed in two different categories: Academia and Industry.

For an academic environment, it should be noted that in the field of power electronics, the simplicity of the new technology is one of the most important features to attract the industry. Any improvements involving a complex or costly solution will not impress the market. On the other hand, a simple design with reliable performance and reduction in size, cost, loss, or maintenance would be appreciated and find a way to the industry. The full-bridge inverters are used widely in the industries due to their simple and well-known structure and operation. Accordingly, the Cascaded H-Bridge (CHB) multilevel topology is popular in very high-power applications due to its simplicity and modularity. The other multilevel inverter topologies in the market that can be mentioned are the Neutral Point Clamp (NPC), Active NPC, and also T3. They all need voltage control for the DC-link capacitors while using a single DC source like a full bridge. As long as the voltage balance technique is integrated into the switching pattern, they would be free of using an external complicated controller as an advantage. The PUC topology was invented as a seven-level inverter with the minimum possible number of switches (six switches only). However, it could not attract the companies to invest in due to lack of simplicity in control. The seven-level topology needs a complex voltage controller which reduces the reliability and easy design/operation for commercialization. As an improvement, the PUC5 has been introduced with a reduced number of levels (generating a five-level voltage), while the voltage controller of the auxiliary capacitor has been integrated into the switching algorithm without using any controller. Moreover, it has a fault-tolerant structure with added reliability while using only two more switches rather than the full bridge. On the other hand, it generates a five-level output voltage with low harmonic contents inherently and features the single-DC-source. All standard controllers for different applications can be applied on this topology with existing implementation procedures on industry rated microcontrollers. All these characteristics made the PUC5 an interesting topology for the companies to go through a new multilevel topology for industrial applications and the power electronics market.

Considering the industrial commercialization, beyond access to capital and well executed marketing, three other often overlooked aspects needed to be properly addressed to achieve successful commercial results. Each aspect will be reviewed in the following paragraphs.

4.5.1 Building a mass-producible product out of a laboratory concept

The PUC5 is one of the most effective ways to perform AC/DC conversion. Having a solid conceptual or laboratory prototype is interesting, but how to mass-produce the device is the real issue.

Laboratory experiments (and this is particularly true in the power electronics space) can easily create misleading impressions about real-world product feasibility. Temperature variation (controlled in the labs, uncontrolled in real life), humidity, vibration, heat dissipation, safety and certification requirements, automated manufacturing and supply chain issues are all elements of the PUC5 initial design that have to be mastered before thinking about commercial production.

Designs involving a PUC5 will also have to manage source electrical noise and power quality issues that are rarely present in a laboratory source.

Can the product (and not only the PUC5) be manufactured without manual processes (like resistance tuning) or nonavailable/limited quantity exotic parts (like GaN /SIC FET)?

These are fundamental issues in designing and building a commercial product that can be manufactured at the best possible cost and whose performance and reliability will not disappoint customers.

4.5.2 Achieving product/market requirement

One way to describe PUC5 is as a function, rather than as a product. The function is to convert AC to DC and vice versa. It is an important function, obviously, but doing the conversion in and of itself does not create any value. It is meeting the needs for power conversion that creates value. And marketing something is all about sensitizing the customer to the value provided. Therefore, the PUC5, in itself, has to be introduced in combination with functionality that customers can recognize and want. On the other side of the coin, the PUC5 must have technology that is sufficiently better to make a difference in the perceived value. It is important to underline that we are not talking here only about monetary value but the full perceived value.

As with any new technology proposed in a critical environment (energy and mobility), introducing even a 10% less expensive but unproven technology is not enough. Customers will perceive an unproven technology as risky/unreliable and will probably conclude that a 10% in cost reduction does not compensate for the risk. The perceived value has to be higher. Form factor, safety, ease of use, warranty are all elements that have to be included for successful commercialization of the technology.

That is why Ossiaco has spent time and money using PUC5 as a core technology to conceive products with a smaller form factor and increased functionality. Lots of time and attention has been dedicated to design aesthetics and user experience and the provision of a solid warranty. All this is not intended to diminish the innovative breakthroughs in the PUC5 concept. It is to showcase the PUC5 in an environment where the end user can appreciate the value that the device brings to them and overcome any perceived risks or fears of new technology.

4.5.3 Keeping the costs/benefits/reliability balance over time

Increasing perceived value can stimulate buyer interest. But cost is usually the main factor triggering the buying process. The PUC5 definitely provides a cost advantage to any product requiring AC<>DC power conversion. But initial selling on cost is only one part of the equation. Customers are expecting a solid warranty (up to 12 years, in case of solar system installers). A longer warranty implies some design choices that lead to a more expensive product. Some customers want to use the product in very difficult environments (+40C to −40C with ice storms like in the North and Midwest USA and Canada, for example). These design adaptations can also lead to price change. Power output is another factor. Successful PUC5 commercialization requires studying and understanding real-world buyer requirements and finding the right balance to propose unbeatable perceived value to each and every customer.

4.6 Conclusions

Multilevel converters have received significant attention for improved power quality. Moreover, replacing isolated DC sources with voltage-controlled capacitors makes this technology much more appealing for the industries due to reduced cost and size. The Packed U-Cells (PUC) inverter is an emerging SDCS-MLI topology, which facilitates the implementation of existing controllers for various applications. This chapter proposed a comprehensive study on the different PUC configurations, modeling, control techniques, applications, and commercialization challenges. It has been shown that the PUC topology combines multiple features of other MLI topologies such as: (1) low impact on the power grid; (2) flexibility in expanding to higher output levels without DC bus extension; (3) ability to offer extended selection of control actions and improvement of the filter bandwidth through switching states redundancy; (4) reliability and low cost due to reduced number of active components; (5) better ride-through capability by means of existing storage capacitors.

Acknowledgment

The authors thank the Ossiaco cofounders who contributed in writing Section 4.5.

References

[1] H. Vahedi, M. Trabelsi, "Single-DC-Source Multilevel Inverters", Springer Nature Switzerland AG, 978-3-030-15252-9, 2019. The final authenticated version is available online at: https://doi.org/10.1007/978-3-030-15253-6.

[2] Y. Ounejjar, K. Al-Haddad, Grégoire, L.-A., Packed U cells multilevel converter topology: theoretical study and experimental validation, IEEE Trans. Ind. Electron. 58 (4) (2011) 1294–1306.

[3] Y. Ounejjar, K. Al-Haddad, L.A. Dessaint, A novel six-band hysteresis control for the packed U cells seven-level converter: experimental validation, IEEE Trans. Ind. Electron. 59 (10) (2012) 3808–3816.

[4] H. Makhamreh, M. Sleiman, O. Kükrer, K. Al-Haddad, Lyapunov-based model predictive control of a PUC7 grid-connected multilevel inverter, IEEE Trans. Ind. Electron. 66 (9) (2019) 7012–7021.

[5] H. Makhamreh, M. Trabelsi, O. Kükrer, H.A. Abu-Rub, A Lyapunov-based model predictive control design with reduced sensors for a PUC7 rectifier, IEEE Trans. Ind. Electron. 68 (2) (2021) 1139–1147.

[6] H. Makhamreh, M. Trabelsi, O. Kükrer, H. Abu-Rub, An effective sliding mode control design for a grid-connected PUC7 multilevel inverter, IEEE Trans. Ind. Electron. 67 (5) (2020) 3717–3725.

[7] H. Makhamreh, M. Trabelsi, O. Kükrer, H. Abu-Rub, A simple sliding mode controller for PUC7 grid-connected inverter using a look-up table, in: IECON 2019—45th Annual Conference of the IEEE Industrial Electronics Society, Lisbon, Portugal, 2019, pp. 3325–3330.

[8] H. Vahedi, P.A. Labbé, K. Al-Haddad, Sensor-less five-level packed U-cell (PUC5) inverter operating in stand-alone and grid-connected modes, IEEE Trans. Ind. Inf. 12 (1) (Feb. 2016) 361–370.

[9] M. Trabelsi, S. Bayhan, K.A. Ghazi, H. Abu-Rub, L. Ben-Brahim, Finite-control-set model predictive control for grid-connected packed-U-cells multilevel inverter, IEEE Trans. Ind. Electron. 63 (11) (Nov. 2016) 7286–7295.

[10] S. Xiao, M. Metry, M. Trabelsi, R.S. Balog, H. Abu-Rub, A model predictive control technique for utility-scale grid connected battery systems using packed U cells multilevel inverter, in: IECON 2016—42nd Annual Conference of the IEEE Industrial Electronics Society, Florence, 2016, pp. 5953–5958.

[11] M. Trabelsi, S. Bayhan, M. Metry, H. Abu-Rub, L. Ben-Brahim, R. Balog, An effective model predictive control for grid connected packed u cells multilevel inverter, in: 2016 IEEE Power and Energy Conference at Illinois (PECI), IL, Urbana, 2016, pp. 1–6.

[12] H. Makhamreh, M. Trabelsi, O. Kükrer, H. Abu-Rub, Model predictive control for a PUC5 based dual output active rectifier, in: 2019 IEEE 13th International Conference on Compatibility, Power Electronics and Power Engineering (CPE-POWERENG), Sonderborg, Denmark, 2019, pp. 1–6.

[13] M. Trabelsi, M. Ghanes, M. Mansouri, S. Bayhan, H. Abu-Rub, An original observer design for reduced sensor control of packed U cells based renewable energy system, Int. J. Hydrog. Energy 42 (28) (2017) 17910–17916.

[14] M. Trabelsi, M. Ghanes, S. Bayhan, H. Abu-Rub, High performance voltage-sensorless model predictive control for grid integration of packed U cells based PV system, in: 2017 19th European Conference on Power Electronics and Applications (EPE'17 ECCE Europe), Warsaw, 2017, pp. 1–8.

[15] B.P. McGrath, D.G. Holmes, Multicarrier PWM strategies for multilevel inverters, IEEE Trans. Ind. Electron. 49 (4) (2002) 858–867.

[16] B. Wu, High-Power Converters and AC Drives, Wiley-IEEE Press, 2006.

[17] K.K. Gupta, P. Bhatnagar, H. Vahedi, K. Al-Haddad, Carrier based PWM for even power distribution in cascaded H-bridge multilevel inverters within single power cycle, in: IECON 2016-42nd Annual Conference of the IEEE Industrial Electronics Society, 2016, pp. 6470–6475.

[18] D.E. Quevedo, R.P. Aguilera, T. Geyer, Predictive control in power electronics and drives: Basic concepts, theory, and methods, in: Advanced and Intelligent Control in Power Electronics and Drives, Springer, 2014, pp. 181–226. 1em plus 0.5em minus 0.4em.

[19] S. Kouro, P. Cortes, R. Vargas, U. Ammann, J. Rodriguez, Model predictive control a simple and powerful method to control power converters, IEEE Trans. Ind. Electron 56 (Jun. 2009).

[20] M. Abarzadeh, H. Vahedi, K. Al-Haddad, Fast sensor-less voltage balancing and capacitor size Reduction in PUC5 converter using novel modulation method, IEEE Trans. Ind. Informat. (2019) 1. vol. Early Access, no.

[21] H. Vahedi, M. Sharifzadeh, K. Al-Haddad, Topology and control analysis of single-DC-source five-level packed U-cell inverter (PUC5), in: IECON 2017-43rd Annual Conference of the IEEE Industrial Electronics Society, 2017, pp. 8691–8696.

[22] S. Arazm, H. Vahedi, K. Al-Haddad, Phase-shift modulation technique for 5-level packed U-cell (PUC5) inverter, in: IEEE 12th International Conference on Compatibility, Power Electronics and Power Engineering (CPE-POWERENG), 2018, pp. 1–6.

[23] S. Arazm, H. Vahedi, K. Al-Haddad, Space vector modulation technique on single phase sensor-less PUC5 inverter and voltage balancing at flying capacitor, in: IECON 2018-44th Annual Conference of the IEEE Industrial Electronics Society, 2018, pp. 4504–4509.

[24] M. Trabelsi, H. Vahedi, H. Komurcugil, H. Abu-Rub, K. Al-Haddad, Low Complexity Model Predictive Control of PUC5 Based Dynamic Voltage Restorer, in: 2018, IEEE 27th International Symposium on Industrial Electronics (ISIE), Cairns, QLD, 2018, pp. 240–245.

CHAPTER 5

Modular multilevel converters

Apparao Dekka[a], Venkata Yaramasu[b], Ricardo Lizana Fuentes[c], and Deepak Ronanki[a]

[a]Department of Electrical Engineering, Lakehead University, Thunder Bay, ON, Canada [b]School of Informatics, Computing, and Cyber Systems (SICCS), Northern Arizona University, Flagstaff, AZ, United States [c]Department of Environment and Energy, Universidad Católica de la Santísima Concepción, Concepcion, Chile

Abbreviations

AC	alternating current
ANPC	active neutral point clamped
APOD	alternate phase-opposition-disposition
CHB	cascaded half-bridge
CSC	current source converter
DC	direct current
DCC	diode clamped converter
DTC	direct torque control
DVR	dynamic voltage restorer
FB	full-bridge
FC	flying capacitor
FOC	field-oriented control
HB	half-bridge
HVAC	high-voltage alternating current
HVDC	high-voltage direct current
IGBT	insulated gate bipolar transistor
LS	level-shifted
MMC	modular multilevel converter
NNPC	nested neutral point clamped
NPC	neutral point clamped
PD	phase-disposition
PI	proportional-integral
POD	phase-opposition-disposition
PS	phase-shifted
PWM	pulse width modulation
RMS	root mean square

SAM	sampled average modulation
SHE	selective harmonic elimination
SM	submodule
SMC	stacked multilevel cell
STATCOM	static synchronous compensator
SVC	static VAr compensator
SVM	space vector modulation
UPQC	unified power quality conditioner
VA	volt-ampere
VOC	voltage-oriented control
VSC	voltage source converter
WF	wind farm

5.1 Introduction

Multilevel converters produce high-quality voltage and current waveforms with low dv/dt and total harmonic distortion and smaller current ripples. The standard multilevel converters such as a diode clamped converter (DCC), flying capacitor (FC) converter, and cascaded H-bridge converter are widely used in medium-voltage, high-power applications [1–3]. The three-level (3 L) DCC, also known as a neutral-point clamped (NPC) converter, is commonly used in many high-power industrial applications. However, the DCC requires a large number of clamping diodes to increase the operating voltage and it is difficult to achieve DC-bus voltage balancing [1]. The FC converter offers simple construction and control [2]; however, large number of flying capacitors reduce the reliability of the converter. Furthermore, the flying capacitors require a large value of capacitance to minimize the capacitor voltage ripples at low frequency of operation [2]. The cascaded H-bridge topology has a modular structure and it is suitable for high-power applications. However, the cascaded H-bridge converter demands a complex phase-shifting transformer with multiple secondary windings to generate the isolated DC source for each H-bridge module [3]. In addition, several advanced multilevel converters including an active neutral-point clamped (ANPC) converter [4], T-type converter [5], nested neutral-point clamped (NNPC) converter [6], and stacked multicell (SMC) converter [7] are developed for high-power applications. The advanced multilevel converters require significant modifications including additional semiconductor devices and passive components to increase their operating voltage and output voltage levels, which is not cost-effective [8]. Furthermore, these converters need to shut down during the device failure and faults, which leads to loss of production and revenue [9].

To overcome the above challenges, a modular multilevel converter (MMC) was first introduced for high-power and high-voltage applications, and it became popular in both academia and industry due to its modularity and scalability features [10]. Furthermore, an MMC allows direct connection to high-voltage grids without a line-frequency transformer, eliminates the isolated DC sources, and is suitable for fault-tolerant operation due to high redundancy [11]. These features attracted a wide range of industry applications including high-voltage direct current (HVDC) transmission systems, medium-voltage motor drives, static synchronous compensator (STATCOM), and offshore wind farms [12]. The first commercial MMC-based HVDC transmission system was designed with 216 submodules (SMs) per arm and installed in the *Transbay Cable* project, USA for a 400 MW power level and 200 kV voltage [13, 14]. The *Curtis-Wright* and *Benshaw* produced the first commercial MMC-based medium-voltage motor drive system with 5–20 SMs per arm [15]. Similarly, *Siemens* developed a STATCOM with 15–200 SMs per arm to reach an operating voltage of 13.9–220 kV [16]. The aforementioned commercial applications were developed with a half-bridge (HB) SM as a building block due to its simple construction and ease of control [17].

Irrespective of SM configuration and the number of SMs per arm, an MMC requires a complex digital control structure to achieve various control objectives including output current control, SM capacitor voltage control, and circulating current control. Among them, the output current control and SM capacitor voltage control are primary objectives and associated with the operation of the MMC [18]. The structure of output current control varies with the application, in which the grid-connected systems are controlled with a voltage-oriented control (VOC) approach [19], whereas the motor drives are controlled with a field-oriented control (FOC) approach [20]. These control methods mainly use synchronous (dq) reference frame-based proportional-integral (PI) regulators to control the currents such that the steady-state and dynamic performance of the load will be greatly improved. On the other hand, the SM capacitor voltage control involves leg voltage control and voltage balancing among the SMs in an arm. The leg voltage control maintains the average DC voltage of each leg at the nominal value of SM capacitor voltage. The SM capacitor voltage balancing can be achieved by either a sorting-based balancing strategy [21] or independent PI regulators [22].

The circulating current control is a secondary objective and it is associated with the size, reliability, and efficiency of an MMC. The circulating currents mainly exist in an MMC due to the difference between the upper and lower arm SM capacitors voltage of a three-phase MMC [23]. These currents mainly consist of negative sequence (second-harmonic) components and increase the Root Mean Square (RMS) value of the arm current [24]. This will further increase the device power losses and affects the converter efficiency. Hence, the reduction of circulating currents is necessary to improve the performance and efficiency of an MMC. The circulating currents can be minimized to a certain extent by selecting a proper size of arm inductor at the design stage of the converter [25]. However, the large value of the arm inductor affects control performance and its dynamics. For the complete elimination of circulating currents, a closed-loop current control in either a synchronous dq-frame or natural abc-frame is required. In the synchronous dq-frame, the dominant second-harmonic component in the circulating current is extracted and minimized by using PI regulators [26]. Alternatively, the resonant regulators are designed in the natural *abc*-frame to minimize the dominant second-harmonic component in the circulating current such that the RMS value of the arm current will be kept below the rated value [27].

In addition, the MMC requires a pulse width modulation (PWM) scheme to generate the gating signals to each SM. These modulation schemes are categorized into high, medium, and fundamental switching frequency PWM schemes. The fundamental switching frequency PWM schemes are employed in MMC-based HVDC applications, where the MMC has a large number of SMs (>200 SMs per arm) to generate an output voltage with the lowest harmonic distortion [28]. These modulation schemes greatly enhance converter efficiency due to low switching losses. The medium and high-switching frequency PWM schemes are employed in MMC-based motor drives, where the MMC has 5–15 SMs per arm [29]. These PWM schemes significantly improve the output power quality and load dynamic performance with a smaller number of SMs.

Considering the present trends, this chapter presents a comprehensive study on the operation, control, and applications of an MMC. The operation of an MMC is analyzed with an HB-SM. In addition, the operation and features of the most popular SMs for an MMC are discussed. An MMC requires digital control to ensure a safe and reliable operation. The generalized synchronous *dq*-frame-based control method is proposed to achieve the control objectives such as SM capacitor voltage control, output current control, and circulating current control. Also, the latest developments in PWM schemes for an MMC are presented. The performance of the control method is analyzed through case studies. Finally, the commercial applications of MMC are summarized in this chapter. Overall, the state-of-the-art and emerging technologies presented in this chapter will refresh the reader's knowledge on this specific topic.

This chapter is organized as follows:

- The fundamental details of MMC and various SM configurations are presented in Section 5.2.
- The overview of the MMC control system and various control objectives are presented in Section 5.3.
- The past and present developments of PWM schemes for an MMC are presented in Section 5.4.
- Sections 5.5 and 5.6 present the philosophy and implementation of SM capacitor voltage and current (output and circulating currents) control methods. Also, the case studies are presented to analyze the performance of the classical control method.
- The commercial applications of an MMC are presented in Section 5.7.
- Section 5.8 provides the concluding remarks of this chapter.

5.2 Fundamentals of a modular multilevel converter

An MMC can reach an operating voltage of medium (2.3–13.8 kV) to high (33–440 kV) and a power capacity of 226 kW to 1000 MW. The operating voltage and power capacity of an MMC are scalable by varying the number of SMs in each

5.2 Fundamentals of a modular multilevel converter

arm. The generalized circuit configuration of a three-phase MMC is shown in Fig. 5.1A. The net DC-bus voltage of an MMC will be generated from an AC source using a rectifier, and it is represented with a simple split DC source of magnitude $V_{dc}/2$ each. The DC-bus is connected to the positive and negative busbar of the upper and lower arms through cables, respectively. These cables are represented by an inductance of L_{dc} and resistance r_{dc}. Each phase of MMC is referred to as a leg, and each leg is further divided into two arms, named as the upper (u) and lower (l) arm. Each arm is composed of a group of SMs in series with an inductor L. This inductor (L) is referred to as an arm inductor and its internal resistance is denoted by r. The midpoint of each leg is referred to as AC terminals and they are connected to the load through a filter inductor L_f with an internal resistance of r_f. The load is represented by resistance R_o.

5.2.1 Principle of operation

The operation of an MMC is analyzed with four HB-SMs in each arm. The connection of SMs in an arm is shown in Fig. 5.1B. The four SM capacitor voltages are rated for v_C each and their respective output voltages are v_{H1}, v_{H2}, v_{H3}, and v_{H4}. The output terminals of each SM are connected in a cascade to form an arm voltage (v_{xy}), where $x \in \{a,b,c\}$ represents the phase and $y \in \{u,l\}$ represents the arm.

The switching states, SM output voltages, and arm voltage are presented in Table 5.1. With four HB-SMs, the arm voltage (v_{xy}) has five voltage levels of

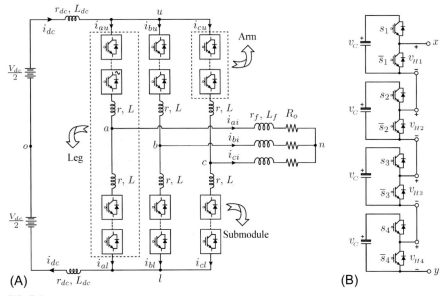

FIG. 5.1

MMC and arm configuration: (A) MMC with a passive load and (B) connection diagram of four HB-SMs in an arm.

$0, v_C, 2v_C, 3v_C,$ and $4v_C$. The voltage level $4v_C$ is generated by turning ON the devices $s_1, s_2, s_3,$ and s_4. Similarly, the voltage level 0 is generated by turning OFF the devices $s_1, s_2, s_3,$ and s_4. Other voltage levels can be generated with multiple switching combinations as presented in Table 5.1. The arm voltage is expressed as

$$\left.\begin{array}{l} v_{xy} = v_{H1} + v_{H2} + v_{H3} + v_{H4} \\ v_{xy} = s_1 v_C + s_2 v_C + s_3 v_C + s_4 v_C \end{array}\right\} \quad (5.1)$$

The arm voltage is equal to the summation of SM output voltages. Each SM output voltage is equal to the product of the SM capacitor voltage and corresponding SM switching state. The SMs in the upper and lower arms of the MMC are controlled to generate a multilevel voltage waveform across the AC load terminals.

5.2.2 Submodule configurations

An MMC can be realized with various SM configurations depending on application requirements such as high-efficiency, DC-side fault current blocking, and smaller SM capacitor voltage ripple and circulating currents [21, 30]. Some of the popular SM configurations are shown in Fig. 5.2.

1) *Half-Bridge Submodule*

Among the several SMs shown in Fig. 5.2, the HB-SM has a simple structure with two semiconductor devices and one DC capacitor and results in a simple control as shown in Fig. 5.2A. During normal operation, only one device will be in the ON state.

Table 5.1 Switching states and voltage levels of an arm.

S_1	S_2	S_3	S_4	v_{H1}	v_{H2}	v_{H3}	v_{H4}	v_{xy}	Voltage level
0	0	0	0	0	0	0	0	0	0
1	0	0	0	v_C	0	0	0	v_C	
0	1	0	0	0	v_C	0	0	v_C	1
0	0	1	0	0	0	v_C	0	v_C	
0	0	0	1	0	0	0	v_C	v_C	
1	1	0	0	v_C	v_C	0	0	$2v_C$	
1	0	1	0	v_C	0	v_C	0	$2v_C$	
1	0	0	1	v_C	0	0	v_C	$2v_C$	2
0	1	1	0	0	v_C	v_C	0	$2v_C$	
0	1	0	1	0	v_C	0	v_C	$2v_C$	
0	0	1	1	0	0	v_C	v_C	$2v_C$	
1	1	1	0	v_C	v_C	v_C	0	$3v_C$	
1	1	0	1	v_C	v_C	0	v_C	$3v_C$	3
1	0	1	1	v_C	0	v_C	v_C	$3v_C$	
0	1	1	1	0	v_C	v_C	v_C	$3v_C$	
1	1	1	1	v_C	v_C	v_C	v_C	$4v_C$	4

5.2 Fundamentals of a modular multilevel converter

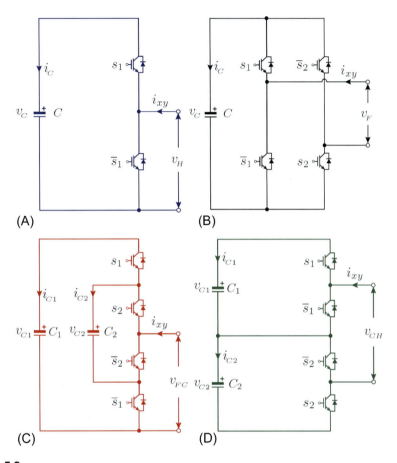

FIG. 5.2

SM configurations: (A) HB-SM, (B) FB-SM, (C) FC-SM, and (D) CHB-SM.

Hence, it has low power losses and high efficiency. The HB-SM generates only two voltage levels of 0 and v_C by using two switching combinations as presented in Table 5.2. The output voltage of HB-SM (v_H) is equal to the product of the SM switching state and DC capacitor voltage, and it is given as

$$v_H = s_1 v_C. \tag{5.2}$$

The DC capacitor voltage either discharge or charge depending on the current direction. From Table 5.2, the DC capacitor current (i_C) is formulated in terms of the switching state and arm current (i_{xy}) as

$$i_C = s_1 i_{xy}. \tag{5.3}$$

When the switching state **1** is applied, the positive direction of the arm current leads to the charging of the SM DC capacitor, whereas the negative direction leads to the

Table 5.2 Switching states of HB-SM.

State	s_1	v_H	$i_{xy} > 0$	$i_{xy} \leq 0$
1	1	v_C	$v_C\uparrow$	$v_C\downarrow$
2	0	0	$v_{C\approx}$	$v_{C\approx}$

discharging of the SM DC capacitor (refer to Table 5.2). For the switching state **2**, there is no change in the SM DC capacitor voltage irrespective of the arm current direction.

2) *Full-Bridge Submodule*

The full-bridge (FB) SM can block the DC-side fault current and it is suitable for MMC-based HVDC applications. Fig. 5.2B shows the FB-SM configuration and it uses four semiconductor devices and one DC capacitor. The FB-SM requires twice the number of semiconductor devices compared with the HB-SM. Hence, it has higher power losses and less efficiency. However, the FB-SM has a smaller DC capacitor voltage ripple compared with an HB-SM. The FB-SM output voltage (v_F) and DC capacitor current (i_C) are given as

$$\left.\begin{array}{l}v_F = (s_1 s_2 - \overline{s_1 s_2})v_c \\ i_C = (s_1 s_2 - \overline{s_1 s_2})i_{xy}\end{array}\right\} \quad (5.4)$$

The FB-SM generates three voltage levels of 0, v_C, and $-v_C$ with four switching states as presented in Table 5.3. The voltage level 0 can be generated by turning ON the device s_1 and turning OFF the device s_2 or turning OFF the device s_1 and turning ON the device s_2. Under this condition, there is no change in the DC capacitor voltage irrespective of the arm current direction as presented in Table 5.3. When the devices s_1 and s_2 are turned ON, the FB-SM generates an output voltage of v_C. In this mode, the DC capacitor charges for the positive arm current direction and discharges for the negative arm current direction. When the devices s_1 and s_2 are turned OFF, the SM generates an output voltage of $-v_C$. This mode of operation is employed to limit the currents during DC-side faults, and it is referred to as the block mode (**BM**) of operation.

Table 5.3 Switching states of FB-SM.

State	s_1	s_2	v_F	$i_{xy} > 0$	$i_{xy} \leq 0$
1	1	1	v_C	$v_C\uparrow$	$v_C\downarrow$
2	1	0	0	$v_{C\approx}$	$v_{C\approx}$
3	0	1	0	$v_{C\approx}$	$v_{C\approx}$
BM	0	0	$-v_C$	$v_C\downarrow$	$v_C\uparrow$

3) Flying Capacitor Submodule

The performance and efficiency of SMs can be improved by using multilevel SMs in place of standard HB and FB-SMs. Furthermore, the use of multilevel SMs leads to a compact design of MMC and improves the controllability due to higher redundancy switching states. The configuration of a three-level (3 L) flying capacitor (FC) SM is shown in Fig. 5.2C, which is realized with four semiconductor devices and two DC capacitors (C_1 and C_2). The voltage of DC capacitor C_1 (v_{C1}) should be twice that of the voltage of DC capacitor C_2 (v_{C2}) so that the FC-SM generates a three-level voltage waveform at the output. The three output voltage levels (0, $v_{C1} - v_{C2}$ or v_{C2} and v_{C1}) can be generated with four switching states as presented in Table 5.4. The output voltage of FC-SM is given as

$$v_{FC} = s_1 v_{C1} + (s_2 - s_1) v_{C2}. \tag{5.5}$$

The current flowing through DC capacitors (C_1 and C_2) is given in terms of the device switching states and arm current as

$$\left. \begin{array}{l} i_{C1} = s_1 i_{xy} \\ i_{C2} = (s_2 - s_1) i_{xy} \end{array} \right\} \tag{5.6}$$

The voltage level 0 will be generated by turning OFF the devices s_1 and s_2. In this mode, the voltage of DC capacitors C_1 and C_2 remains constant irrespective of the arm current direction. When the devices s_1 is ON and s_2 is OFF, the voltage level $v_{C1} - v_{C2}$ is obtained. In this mode, the voltage of DC capacitor C_1 (v_{C1}) increases while the voltage of DC capacitor C_2 (v_{C2}) decreases for the positive direction of arm current and vice versa. This voltage level is also generated by using the voltage of DC capacitor C_2 alone, and the corresponding switching combination is s_1-OFF and s_2-ON. In this mode, the voltage of DC capacitor C_2 would be affected by the current direction, whereas there would be no change in the voltage of DC capacitor C_1. The voltage level v_{C1} is generated by turning ON the devices s_1 and s_2. In this mode, the voltage of DC capacitor C_1 increases for the positive current direction and decreases for the negative current direction. There would be no change in the voltage of DC capacitor C_2.

The 3 L FC-SM can be operated as an HB-SM by applying identical gating signals to the switching devices s_1 and s_2. Under this condition, the DC capacitor C_2 is completely bypassed from the operation and DC capacitor C_1 is used to generate the

Table 5.4 Switching states of FC-SM.

State	s_1	s_2	v_{FC}	$i_{xy} > 0$	$i_{xy} \leq 0$
1	1	1	v_{C1}	$v_{C1} \uparrow$, $v_{C2} \approx$	$v_{C1} \downarrow$, $v_{C2} \approx$
2	0	1	v_{C1}	$v_{C1} \approx$, $v_{C2} \uparrow$	$v_{C1} \approx$, $v_{C2} \downarrow$
3	1	0	$v_{C1} - v_{C2}$	$v_{C1} \uparrow$, $v_{C2} \downarrow$	$v_{C1} \downarrow$, $v_{C2} \uparrow$
4	0	0	0	$v_{C1} \approx$, $v_{C2} \approx$	$v_{C1} \approx$, $v_{C2} \approx$

required output voltage. The mathematical representation of SM output voltage is given as

$$v_H = s_1 s_2 v_{C1}. \tag{5.7}$$

4) *Cascaded Half-Bridge Submodule*

Another possible multilevel SM is the cascade connection of two half-bridge SMs as shown in Fig. 5.2D. This SM is commercialized in MMC-based motor drives [15]. The cascaded half-bridge (CHB) SM consists of four semiconductor devices and two DC capacitors. The resultant circuit generates a three-level voltage waveform at the output. The two DC capacitor voltages are regulated at identical values ($v_{C1} = v_{C2} = v_C$) to obtain a three-level output voltage with an equal voltage step of v_C. The switching states corresponding to the three-level operation of the CHB-SM are presented in Table 5.5. The output voltage of CHB-SM is expressed as

$$v_{CH} = s_1 v_{C1} + s_2 v_{C2}. \tag{5.8}$$

When the devices s_1 and s_2 are turned ON, the SM generates a voltage level of $v_{C1} + v_{C2}$. In this mode, the voltage of DC capacitor C_1 and C_2 increases for the positive current direction and decreases for the negative current direction. The switching combinations s_1-ON and s_2-OFF or s_1-OFF and s_2-ON are used to generate the voltage levels v_{C1} or v_{C2}, respectively. When the devices s_1 and s_2 are turned OFF, the SM generates 0 voltage level at the output, and there is no change in the DC capacitors voltage.

5.3 Classical control methods

The closed-loop control method enables a safe, reliable, and high-performance operation for an MMC. The control of an MMC involves multiple control objectives, which are categorized into primary and secondary objectives as shown in Fig. 5.3. The output current and SM capacitor voltage control are primary objectives associated with the operation of an MMC. On the other hand, the circulating current control is a secondary objective associated with the efficiency and reliability of an MMC. These control objectives are achieved with a classical control method which involves a reference frame theory, PI-regulators, and pulse width modulator [31].

Table 5.5 Switching states of CHB-SM.

State	s_1	s_2	v_{CH}	$i_{xy} > 0$	$i_{xy} \leq 0$
1	1	1	$v_{C1} + v_{C2}$	$v_{C1}\uparrow, v_{C2}\uparrow$	$v_{C1}\downarrow, v_{C2}\downarrow$
2	1	0	v_{C1}	$v_{C1}\uparrow, v_{C2}\approx$	$v_{C1}\downarrow, v_{C2}\approx$
3	0	1	v_{C2}	$v_{C1}\approx, v_{C2}\uparrow$	$v_{C1}\approx, v_{C2}\downarrow$
4	0	0	0	$v_{C1}\approx, v_{C2}\approx$	$v_{C1}\approx, v_{C2}\approx$

5.3 Classical control methods

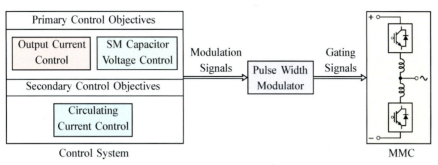

FIG. 5.3

Control objectives for an MMC.

The generalized structure of the centralized classical control method is shown in Fig. 5.4. In this method, independent PI regulators in the dq-reference frame are used to control the output and circulating currents. On the other hand, the SM capacitor voltage control consists of two sub-controls named as the leg voltage control and voltage balancing strategy [22, 32]. The leg voltage control maintains the average DC voltage of each leg at the nominal value of the SM capacitor voltage (v_C). The leg voltage control generates a compensating signal v_{dct}^* [33], which is added to the upper and lower arm modulation signals as shown in Fig. 5.4. Similarly, the output and circulating current controller generates control signals v_{xi}^* and v_{xz}^*, respectively [34]. These control signals are used to form the reference modulation signals for the upper and lower arms. These modulation signals are mathematically represented as.

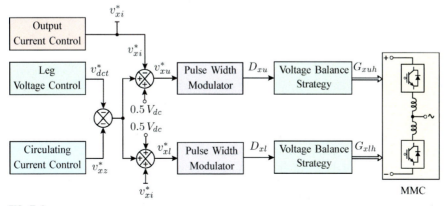

FIG. 5.4

Block diagram of the centralized control method for an MMC.

$$\left. \begin{array}{l} v_{xu}^* = \dfrac{V_{dc}}{2} - v_{xi}^* - v_{xz}^* - v_{dct}^* \\ v_{xl}^* = \dfrac{V_{dc}}{2} + v_{xi}^* - v_{xz}^* - v_{dct}^* \end{array} \right\} \qquad (5.9)$$

where v_{xu}^* and v_{xl}^* represent the upper and lower arm reference modulation signals.

The upper and lower arm modulation signals are given to the modulation stage to generate the gating signals. The modulation stage provides the voltage level information, which needs to be generated at the converter output AC terminals. The voltage level information also represents the required number of ON state SMs in each arm. This information is given to the voltage balancing strategy and the balancing strategy selects ON state SMs from N submodules in an arm based on the voltage level, arm current direction, and instantaneous value of the SM capacitor voltage [35]. By doing so, the converter generates the required output voltage, while maintaining each SM capacitor's voltage at its nominal value.

5.4 Pulse width modulation schemes

The pulse width modulation (PWM) schemes are employed to achieve a wide range of objectives such as capacitor voltage control, common-mode voltage minimization, reduction of device switching frequency and power losses, and minimization of output current ripple. Depending on the switching frequency, the PWM schemes for an MMC are categorized into high, medium, and fundamental switching frequency PWM schemes as shown in Fig. 5.5. The multi-carrier PWM schemes such as phase-shifted and level-shifted carrier PWM schemes belong to the high-switching frequency category [36]. These modulation schemes are applied to the MMC-based motor drive system. In the phase-shifted carrier PWM, the carrier signals are disposed of with a phase shift of $360°/2N$ (with interleaving) or $360°/N$ (without interleaving) [37, 38]. Among them, the phase-shifted carrier PWM with interleaving generates an output voltage with the lowest harmonic distortion compared with no interleaving between the upper and lower arm carrier signals. On the other hand, the level-shifted carrier PWM is further categorized into phase-disposition (PD), phase-opposition disposition (POD), and alternate phase-opposition-disposition (APOD) PWM schemes [39]. Among them, the PD-PWM has the lowest voltage harmonic distortion and maintains equal device switching frequency compared with the POD and APOD PWM schemes. However, the PD-PWM generates higher output current ripple and circulating currents compared with the phase-shifted carrier PWM scheme [29].

The sampled average modulation (SAM) and space vector modulation (SVM) schemes belong to the medium-switching frequency category. The SAM approach is a per-phase approach and directly controls the phase voltages of MMC; it generates line-to-line voltages implicitly [40]. On the other hand, the SVM approach directly controls line-to-line voltages and generates phase voltages implicitly [41]. The three-phase equivalent of SAM is similar to the SVM, except that the zero

5.4 Pulse width modulation schemes

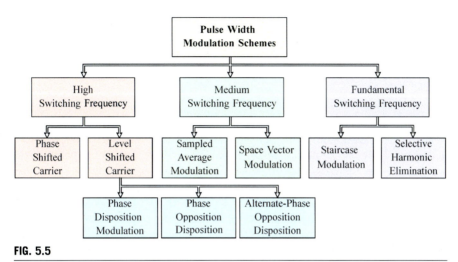

FIG. 5.5

Pulse width modulation schemes for an MMC.

vectors at the start and end of the switching sequence are not equally distributed. This approach leads to a slightly higher harmonic distortion compared with SVM, but it has less computational complexity and is suitable for MMC-based HVDC applications [42].

The staircase modulation and selective harmonic elimination (SHE) schemes belong to the fundamental switching frequency category. In staircase modulation, MMC SMs are selected based on the nearest voltage level and switched at the fundamental frequency [43]. So, the SM switching power losses are very low. Furthermore, it is suitable for MMC-based HVDC applications. On the other hand, the SHE schemes involve an off-line calculation of switching angles to eliminate the dominant harmonic components from the output voltage. The number of switching angles increases with the number of SMs [44]. Hence, it is difficult to apply for MMCs with a large number of SMs. Furthermore, the SHE schemes have poor dynamic performance compared with the staircase modulation scheme. Overall, the phase-shifted carrier PWM and staircase modulation schemes are widely used in various commercial applications of MMC. The design and implementation of these modulation schemes are discussed in the following sections.

5.4.1 Phase-shifted carrier modulation scheme

The reference modulation signals and triangular carrier signals are required to implement the phase-shifted carrier PWM scheme. According to Eq. (5.9), the upper and lower arm reference modulation signals require the following voltage components: phase modulation signals (v_{xi}^*), circulating current compensation signal (v_{xz}^*), and leg voltage control signal (v_{dct}^*). Depending on the application, the phase modulation signals are generated through either the open-loop or closed-loop output current control loops. The voltage-oriented control (VOC) in grid-connected systems and the field-oriented control (FOC) in motor drives are commonly employed to generate the reference phase modulation signals. The reference phase modulation signal is defined as

$$v_{xi}^* = m_a \times \frac{V_{dc}}{2} \sin(\omega_o t + \theta_x) \tag{5.10}$$

where m_a is the amplitude modulation index with the range between 0 and 1. ω_o is the fundamental angular frequency and $\theta_x \in \{0, -\frac{2\pi}{3}, -\frac{4\pi}{3}\}$ represents the phase angle between the three-phase signals in radians.

The phase modulation signals are equally divided and added to the upper and lower arm reference modulation signals [see Eq. (5.9)]. The circulating current and leg voltage control signals are obtained from their respective controllers and used to form the resultant upper and lower arm reference modulation signals.

Along with the reference modulation signals, the phase-shifted carrier PWM requires triangular carrier signals. The triangular carrier signals are arranged in horizontal disposition with a phase-shift between the adjacent carrier signals. These carrier signals have the same frequency and peak-to-peak amplitude. The phase-shifted carrier PWM requires a total of 2 N triangular carrier signals per phase to control an MMC with N SMs per arm. In this approach, the same triangular carrier signals are used to control the upper and lower arm SMs. This approach is referred to as non-interleaved phase-shifted carrier PWM, and the phase angle between adjacent carrier signals is 360°/N. With this approach, MMC generates N + 1 voltage levels at the output [37, 38]. Also, the upper and lower arm SMs can be controlled with different triangular carrier signals. This approach is referred to as an interleaved phase-shifted carrier PWM. In this approach, the phase angle between adjacent carrier signals is 360°/2N and it can generate 2 N + 1 voltage levels at the output [37, 38].

5.4.2 Staircase modulation scheme

Unlike carrier PWM schemes, the staircase modulation scheme does not require any carrier signals in its implementation. Hence, it is easy to apply for MMC-based HVDC systems. In this approach, the voltage levels and duty cycles are directly obtained from the reference phase modulation signals [43]. Due to the fundamental switching frequency operation, the staircase modulation scheme generates output voltage with high harmonic distortion if there are a lesser number of SMs in each arm of an MMC. The approximated normalized upper and lower arm modulation signals (without circulating current and leg voltage control signals) are given as

$$\left.\begin{aligned} v_{xu}^n &= \frac{N}{2} \times [1 - m_a \sin(\omega_o t + \theta_x)] \\ v_{xl}^n &= \frac{N}{2} \times [1 + m_a \sin(\omega_o t + \theta_x)], \end{aligned}\right\} \tag{5.11}$$

where v_{xu}^n and v_{xl}^n represent the normalized upper and lower arm voltages.

The normalized voltage waveform has steps in the range of 0 to $N+1$ and represents the number of ON state SMs in an arm. The required voltage level at each sampling interval can be obtained from

$$\left.\begin{array}{c} m_{xu}^n = round\left(v_{xu}^n\right) \\ m_{xl}^n = round\left(v_{xl}^n\right), \end{array}\right\} \tag{5.12}$$

where m_{xu}^n and m_{xl}^n represent the nearest voltage level of the upper and lower arm instantaneous voltage.

The nearest voltage level information is used along with the current direction and instantaneous value of SM capacitor voltage in the selection process of ON state SMs out of N SMs in an arm.

5.5 Submodule capacitor voltage control

An MMC consists of multiple SMs in each arm, and each SM consists of flying capacitors. These flying capacitors voltage must be controlled at their nominal value such that the MMC can generate a multilevel voltage waveform across the load terminals. To regulate the flying capacitors voltage, a simple control method is proposed. The SM capacitor voltage control of MMC has two main functions: (i) maintaining the average voltage of each leg to a nominal value of SM capacitor voltage, and (ii) voltage balancing among the SMs within the arm. These two functions are achieved with leg voltage control and voltage balancing strategy, respectively [21, 33].

5.5.1 Leg voltage control

The main objective of leg voltage control is to maintain the average value of upper and lower arm SM capacitor voltages in each leg at a value of v_C. This objective is achieved by controlling the DC current component in the arm current. The control block diagram of leg voltage control is shown in Fig. 5.6. The leg voltage control consists of an outer voltage control loop and an inner current control loop [21]. The outer voltage control loop minimizes the difference between the reference and measured average leg voltage. This control loop provides reference DC current component command (i_{xd}^*). The reference DC current component command is given by

$$i_{xd}^* = k_{pv}\left(V_C^* - v_{Cx}\right) + k_{iv}\int\left(V_C^* - v_{Cx}\right)dt, \tag{5.13}$$

where k_{pv} and k_{iv} are the proportional and integral gain of the voltage control loop, respectively, and V_C^* is the reference value of average leg voltage.

The measured average leg voltage (v_{Cx}) is given by

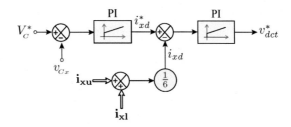

FIG. 5.6

Block diagram of MMC leg voltage control.

$$v_{Cx} = \frac{1}{2N}\left[\sum_{h=1}^{N} v_{Cuh}^x + \sum_{h=1}^{N} v_{Clh}^x\right], \qquad (5.14)$$

where v_{Cuh}^x and v_{Clh}^x are the SM capacitor voltages in the upper and lower arms of phase-x.

The inner current control loop minimizes the difference between the reference and measured DC current component. This control loop provides a voltage command of

$$v_{dct}^* = k_{pi}(i_{xd}^* - i_{xd}) + k_{ii}\int (i_{xd}^* - i_{xd})dt, \qquad (5.15)$$

where k_{pi} and k_{ii} are the proportional and integral gain of the current control loop, respectively.

The actual DC current component is calculated from the measured three-phase upper and lower arm currents, and it is given by

$$i_{xd} = \frac{i_{xu} + i_{xl}}{6}, \qquad (5.16)$$

where $\mathbf{i_{xu}} \in \{i_{au}, i_{bu}, i_{cu}\}$ and $\mathbf{i_{xl}} \in \{i_{al}, i_{bl}, i_{cl}\}$.

5.5.2 Voltage balancing strategy

The main function of the voltage balancing strategy is to equally distribute the net DC-bus voltage (V_{dc}) between the SMs in each arm. This objective is achieved with an algorithm that uses logical functions and comparison logic or sorting technique [21, 32]. The principle of the voltage balancing strategy is to regulate the charging and discharging of SM capacitors based on the instantaneous value of capacitors voltage and arm current direction. For negative current direction, the SM capacitors with the highest voltage are inserted in an arm such that these SM capacitors will discharge leading to a reduction in voltage close to the nominal value and vice versa. The steps involved in the implementation of the voltage balancing strategy are shown in Fig. 5.7A [42].

5.5 Submodule capacitor voltage control

- SM capacitor voltages ($v_{Cyh}^x \in \{v_{Cy1}^x, v_{Cy2}^x, \ldots, v_{CyN}^x\}$) in each arm are measured, and their voltage magnitudes are given to the comparison logic as shown in Fig. 5.7B.
- Each SM capacitor voltage is compared with other SM capacitor voltages in the same arm using the comparison logic. The resultant output of each comparison is added together to form an index number (VI_h). The lowest index number is assigned to the SM with the highest capacitor voltage and vice versa.
- Obtain the current direction (D) from the measured arm current (i_{xy}). The value of $D = 1$ represents the positive current direction and $D = 0$ represents the negative current direction.
- The SM index numbers are arranged in either the ascending or descending order based on the arm current direction by using the following expression,

$$AI_h = VI_h \times D + (N - 1 - VI_h) \times (1 - D). \tag{5.17}$$

By using Eq. (5.17), the SMs with the highest capacitor voltage are inserted in the arm for the negative current direction and are discharged (decreasing their voltage), and vice versa.

- From the modulation stage, obtain the number of SMs needed to be inserted in an arm (D_{xy}).
- Compare the actual index number of each SM (AI_h) with the reference index number ($N - D_{xy}$) to generate the $INSERT = 1$ or $BYPASS = 0$ states for each SM (S_{xyh}).

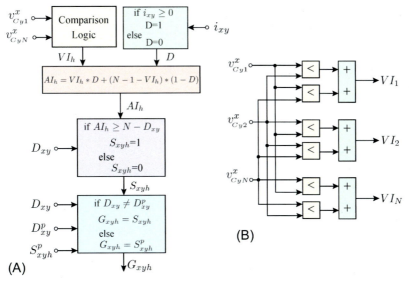

FIG. 5.7

Voltage balance strategy for an MMC: (A) flowchart of balancing strategy and (B) comparison logic.

- The inserted number of SMs in the present control cycle (D_{xy}) is compared with the previous control cycle (D_{xy}^p). If there is a difference in D_{xy} and D_{xy}^p, then the new gating signals (i.e., $G_{xyh} = S_{xyh}$) are applied, or else the gating signals in the previous control cycle are maintained (i.e., $G_{xyh} = S_{xyh}^p$). This approach eliminates the unwanted SM switchings and ensures the SM switching frequency is equal to the triangular carrier frequency.

The dynamic performance of SM capacitor voltage control is verified on an MMC with a rated apparent power $S_o = 2$ MVA, output line-to-line voltage $V_o = 6$ kV, rated output frequency $f_o = 60$ Hz, rated power factor (PF) = 0.85 (lag), rated DC-bus voltage $V_{dc} = 10$ kV, number of SMs per arm $N = 4$, arm inductance $L = 5$ mH, and SM capacitance $C = 2200$ µF. The voltage balancing strategy is enabled from $t = 0.0$ s to $t = 0.1$ s. Fig. 5.8A shows that the leg voltage control maintains the average voltage of leg-a at 1 pu (2.5 kV). Each arm of the MMC consists of four SMs, and each SM capacitor voltage in the upper and lower arms is regulated at 1 pu (2.5 kV) [Figs. 5.8B and C]. From $t = 0.1$ s to $t = 0.15$ s, the voltage balancing strategy is disabled. The SM capacitors voltage in the upper and lower arms are diverging from their rated value (1 pu). This will affect the output power quality, device voltage stress, and converter reliability. In addition, the average leg voltage is deviating from its rated value of 1 pu due to the difference in SM capacitors voltage in the upper and lower arms. At $t = 0.15$ s, the balancing strategy is enabled, which ensures all the SM capacitors voltage are maintained at their rated value.

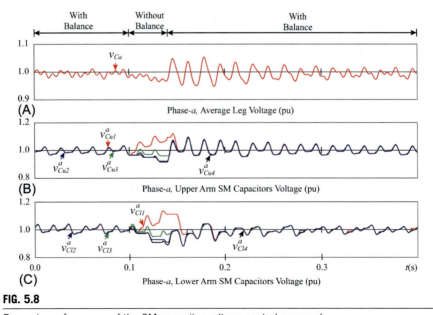

FIG. 5.8

Dynamic performance of the SM capacitor voltage control approach.

5.6 Current control

The current control of the MMC needs to ensure perfect tracking of three-phase output currents to their reference for better load performance. In addition, the circulating currents should be minimized to improve the efficiency and reliability of the MMC. Overall, the MMC requires a high-performance current control to achieve safe and reliable operation. The output and circulating current controls are implemented in a synchronous *dq*-frame, where the time-varying control signals are transformed into DC signals, and these DC signals are regulated by using a simple PI-regulator [19]. The output and circulating currents are controlled with independent PI regulators and their details are discussed for an MMC with a resistive-inductive (RL) load.

5.6.1 Output current control

The per-phase equivalent circuit of an MMC is shown in Fig. 5.9, in which the upper and lower arm SM output voltages are modeled as a single voltage source of v_{xu} and v_{xl}, respectively. The upper and lower arm currents consist of DC current (i_{dc}), circulating current (i_{xz}), and output current (i_{xi}) components [37]. The arm currents are defined as

$$\left. \begin{aligned} i_{xu} &= \frac{1}{3}i_{dc} + i_{xz} + \frac{1}{2}i_{xi} \\ i_{xl} &= \frac{1}{3}i_{dc} + i_{xz} - \frac{1}{2}i_{xi} \end{aligned} \right\} \quad (5.18)$$

The upper and lower arm currents consist of an equal amount of output current components, which is obtained by subtracting the lower arm current from the upper arm current as

$$i_{xi} = i_{xu} - i_{xl}. \quad (5.19)$$

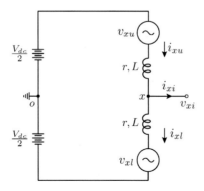

FIG. 5.9

Per-phase equivalent circuit of MMC.

The output currents are controlled with *dq*-frame-based PI regulators. The control block diagram of output current control for an MMC with *RL*-load is shown in Fig. 5.10. The three-phase reference (i_{ai}^*, i_{bi}^*, i_{ci}^*) and measured output currents (i_{ai}, i_{bi}, i_{ci}) are converted into a *dq*-frame rotating at a speed equal to the rated frequency of the load (ω_o) [45]. The *abc* to *dq* transformation is given by

$$\begin{bmatrix} f_d \\ f_q \end{bmatrix} = \frac{2}{3} \begin{bmatrix} \cos\theta & \cos\left(\theta - \frac{2\pi}{3}\right) & \cos\left(\theta - \frac{4\pi}{3}\right) \\ -\sin\theta & -\sin\left(\theta - \frac{2\pi}{3}\right) & -\sin\left(\theta - \frac{4\pi}{3}\right) \end{bmatrix} \begin{bmatrix} f_a \\ f_b \\ f_c \end{bmatrix}, \quad (5.20)$$

where f_d, f_q are *d*- and *q*-axis components (current or voltage), respectively, f_a, f_b, f_c are three-phase components (current or voltage) in the *abc*-reference frame, and θ is the output phase angle.

The reference and actual *d*- and *q*-axis current components are compared, and their difference leads to current errors. The *d*- and *q*-axis current component errors Δi_{di} and Δi_{qi} are given to the PI regulators. The current PI-regulator minimizes the current errors and generates reference *d*- and *q*-axis voltage commands v_{di}^* and v_{qi}^*, respectively. The reference *dq*-frame voltage commands are transformed into the *abc*-frame voltages by using the transformation matrix given in Eq. (5.21). The reference *abc*-frame voltages are added to arm modulation signals to control an MMC.

$$\begin{bmatrix} f_a \\ f_b \\ f_c \end{bmatrix} = \begin{bmatrix} \cos\theta & -\sin\theta \\ \cos\left(\theta - \frac{2\pi}{3}\right) & -\sin\left(\theta - \frac{2\pi}{3}\right) \\ \cos\left(\theta - \frac{4\pi}{3}\right) & -\sin\left(\theta - \frac{4\pi}{3}\right) \end{bmatrix} \begin{bmatrix} f_d \\ f_q \end{bmatrix}. \quad (5.21)$$

The dynamic performance of output current control is verified on an MMC with a rated apparent power $S_o = 2$ MVA, output line-to-line voltage $V_o = 6$ kV, rated output frequency $f_o = 60$ Hz, rated power factor (PF) = 0.85 (lag), rated DC-bus voltage

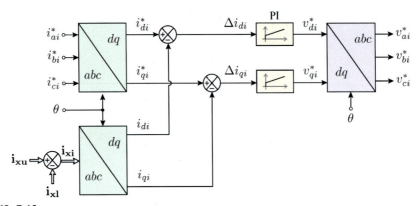

FIG. 5.10

Output current control in the synchronous *dq*-frame.

5.6 Current control

$V_{dc} = 10$ kV, number of SMs per arm $N = 4$, arm inductance $L = 5$ mH, and SM capacitance $C = 2200$ μF. The dynamic performance of output current control in the dq-frame with a step change in the reference output current magnitude is shown in Fig. 5.11. From $t = 0.0$ s to $t = 0.05$ s, the reference output current magnitude and its frequency are set to $I^* = 0.4$ pu and $f_o = 60$ Hz, respectively. The i_{di} and i_{qi} current components are regulated at $+0.34$ pu and -0.21 pu corresponding to the 0.85 (lag) power factor, respectively [Figs. 5.11A and B]. The three-phase output currents have a peak value of 0.4 pu as shown in Fig. 5.11C. At $t = 0.05$ s, the reference current magnitude is changed from $I^* = 0.4$ pu to $I^* = 0.8$ pu. The reference d- and q-axis current components are also changed from $i_{di}^* = +0.34$ pu to $i_{di}^* = +0.64$ pu, and $i_{qi}^* = -0.21$ pu to $i_{qi}^* = -0.42$ pu, respectively. However, the PI regulators are perfectly controlling the converter to generate the given reference currents at the output. The peak value of measured output currents in the abc-reference frame perfectly matches with the reference current magnitude.

5.6.2 Circulating current control

The upper and lower arm currents given in Eq. (5.18) are added together, which results in a common-mode current component flowing through each leg of an MMC. The common-mode current component (i_{xzf}) consists of a DC current component and circulating current component, and it is given by

$$i_{xzf} = \frac{1}{2}(i_{xu} + i_{xl}) = \frac{1}{3}i_{dc} + i_{xz}. \tag{5.22}$$

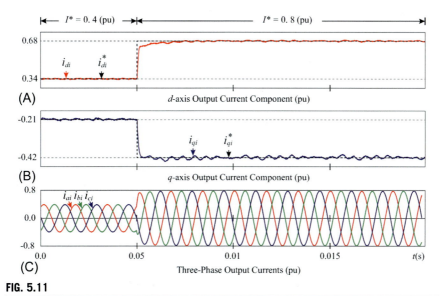

FIG. 5.11

Dynamic performance of output current control.

From the equivalent circuit shown in Fig. 5.9, the upper and lower arm voltages are given by

$$\left. \begin{array}{l} v_{xu} = \dfrac{V_{dc}}{2} - v_{xi} - L\dfrac{di_{xu}}{dt} - ri_{xu} \\ v_{xl} = \dfrac{V_{dc}}{2} + v_{xi} - L\dfrac{di_{xl}}{dt} - ri_{xl}, \end{array} \right\} \quad (5.23)$$

where v_{xu}, v_{xl} are the upper and lower arm voltages, respectively, and v_{xi} is the output voltage.

From the models in Eqs (5.22) and (5.23), the voltage drop across the arm inductor due to the common-mode current component is given by [37]

$$2L\frac{di_{xzf}}{dt} + 2ri_{xzf} = V_{dc} - (v_{xu} + v_{xl}). \quad (5.24)$$

From the models in Eqs (5.22) and (5.24), the voltage drop across the arm inductor due to circulating current is given by

$$L\frac{di_{xz}}{dt} + ri_{xz} = \frac{V_{dc}}{2} - \frac{1}{2}(v_{xu} + v_{xl}) - r\frac{i_{dc}}{3} \quad (5.25)$$

The DC current component (i_{dc}) is controlled through leg voltage control to achieve a power balance between the input and output of the MMC. To simplify the analysis, the voltage drop due to the DC current component ($r\frac{i_{dc}}{3}$) is neglected. The simplified circulating current model is given by [37]

$$L\frac{di_{xz}}{dt} + ri_{xz} = \underbrace{\frac{V_{dc}}{2} - \frac{1}{2}(v_{xu} + v_{xl})}_{v_{xz}}, \quad (5.26)$$

From Eq. (5.26), the three-phase circulating current model in the *abc*-reference frame is calculated as

$$\begin{bmatrix} v_{az} \\ v_{bz} \\ v_{cz} \end{bmatrix} = L\frac{d}{dt}\begin{bmatrix} i_{az} \\ i_{bz} \\ i_{cz} \end{bmatrix} + r\begin{bmatrix} i_{az} \\ i_{bz} \\ i_{cz} \end{bmatrix} \quad (5.27)$$

The circulating current mainly consists of even-order harmonic components in which the second-order harmonic component is the dominant component. These harmonic components are time-varying in nature and difficult to control using PI regulators without any steady-state error. Hence, these time-varying signals are converted into DC signals by using the transformation matrix given in Eq. (5.20). In this study, the controller is designed to eliminate the second-order harmonic component from the circulating current. Therefore, the circulating current model in a synchronous-*dq* frame rotating at $-2\omega_o$ frequency is given as [37]

$$\begin{bmatrix} v_{dz} \\ v_{qz} \end{bmatrix} = L\frac{d}{dt}\begin{bmatrix} i_{dz} \\ i_{qz} \end{bmatrix} + \begin{bmatrix} 0 & -2\omega_o L \\ 2\omega_o L & 0 \end{bmatrix}\begin{bmatrix} i_{dz} \\ i_{qz} \end{bmatrix} + r\begin{bmatrix} i_{dz} \\ i_{qz} \end{bmatrix}. \quad (5.28)$$

Based on the mathematical model given in Eq. (5.28), the circulating current control is designed, and its block diagram is shown in Fig. 5.12. The actual circulating currents are estimated from the measured arm currents. These currents are converted into dq-frame resulting in i_{dz} and i_{qz} current components. The reference d- and q-axis circulating currents (i_{dz}^* and i_{qz}^*) are set to zero. The comparison between the reference and actual circulating current gives the current errors Δi_{dz} and Δi_{qz}. These current errors are minimized by using PI regulators. The d- and q-axis circulating current control loops are decoupled by adding the induced speed voltages in the arm inductor to the current control loops. The PI-regulator generates reference dq-frame voltage signals v_{dz}^* and v_{qz}^*. The reference dq-frame voltages are converted back to the abc-reference frame by using the transformation matrix given in Eq. (5.21). The three-phase reference voltage signals (v_{az}^*, v_{bz}^*, v_{cz}^*) are used to form the arm modulation signals to control an MMC.

The performance of circulating current control is verified on an MMC with a rated apparent power $S_o = 2$ MVA, output line-to-line voltage $V_o = 6$ kV, rated output frequency $f_o = 60$ Hz, rated power factor (PF) $= 0.85$ (lag), rated DC-bus voltage $V_{dc} = 10$ kV, number of SMs per arm $N = 4$, arm inductance $L = 5$ mH, and SM capacitance $C = 2200$ μF. The performance of the circulating current controller in the dq-frame is shown in Fig. 5.13. Initially, the circulating current controller is disabled, and the d- and q-axis circulating current components have a peak value of 0.4 pu (Fig. 5.13A). The circulating current in the abc-reference frame (i_{az}) mainly consists of a second-order harmonic component as shown in Fig. 5.13B. These harmonic components increase the peak and RMS value of the arm current flowing through the converter legs (Fig. 5.13C). At $t = 0.03$ s, the circulating current controller is activated by setting the i_{dz}^* and i_{qz}^* values to zero. The PI-regulator effectively forces the actual i_{dz} and i_{qz} magnitudes close to zero (Fig. 5.13A). Also, the reduction of a second-harmonic current component leads to a smaller circulating current in the arm (Fig. 5.13B). The second-order harmonic component in the circulating current is also minimized significantly, which further reduces the converter power losses and

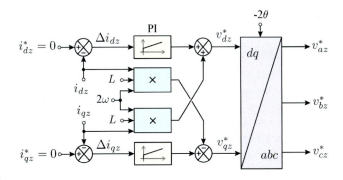

FIG. 5.12

Circulating current control in the synchronous-dq-frame.

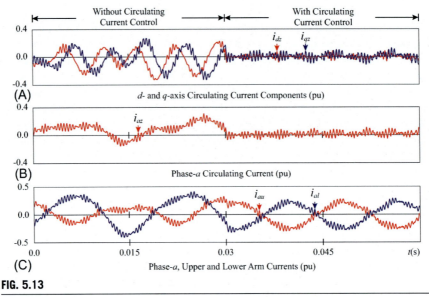

FIG. 5.13

Performance of circulating current control in the *dq*-frame.

improves efficiency. Moreover, the upper and lower arm currents are symmetrical and sinusoidal in nature (Fig. 5.13C).

5.7 Applications

MMCs are commercialized in the form of standard and customized products for high-power applications. An MMC is used in several industrial applications such as HVDC transmission systems, motor drives, offshore wind farms (WFs), and power quality improvement [46–50]. The summary of commercially available MMC-based technologies is presented in the following sections.

5.7.1 HVDC transmission systems

HVDC transmission systems are cost-effective for bulk power transfer over long distances with low power losses. These systems have higher controllability and stability and require a less expensive busbar structure compared with high-voltage alternating current (HVAC) transmission systems. HVDC systems are employed to interconnect two asynchronous or synchronous AC power systems using current source converters (CSCs) or voltage source converters (VSCs). The early HVDC technology uses a thyristor-based CSC to handle higher voltage and power capacity. However, the CSC-based HVDC requires harmonic filters to improve the power factor and AC power quality [51]. Currently, the VSC-based HVDC technology is widely used due

5.7 Applications

to the independent control of active and reactive powers, and they have the ability to operate with weak AC grids [52].

Currently, MMC-based HVDC technology is commercially available for bulk power transfer. An MMC can handle a high-voltage operation with low-cost and low-voltage semiconductor devices, and significantly increases the output voltage levels, enabling a reduction in the filter size and power losses [53]. The back-to-back configuration of the MMC-based HVDC transmission system is shown in Fig. 5.14. The MMC-I and MMC-II are connected to grid-I and grid-II through a transformer, respectively. The DC-sides of MMC-I and MMC-II are connected through a DC cable several hundreds of kilometers long. In HVDC applications, an MMC is designed with a larger number of SMs of 200–400 SMs per arm to reach an operating voltage of ± 320 kV.

A list of commercial MMC-based HVDC projects designed by leading industrial manufacturers is presented in Table 5.6. These projects are designed with four types of HVDC technologies named as *HVDC Plus*, *HVDC Light Gen.4*, *HVDC Flexible*,

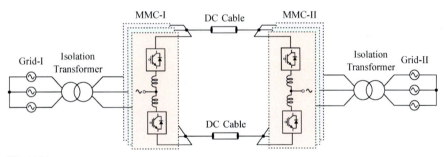

FIG. 5.14

MMC-based HVDC transmission system.

Table 5.6 Market overview of MMC-based HDVC systems [11, 18, 54].

Project name	Power (MW)	Voltage ($\pm$ kV)	DC cable (km)	Company	HVDC Tech.
Trans Bay	400	200	85	Siemens	HVDC Plus
INELFE	2 × 1000	320	65	Siemens	HVDC Plus
Sydvastlanken	2 × 720	300	260	Alstom	HVDC MaxSine
Piemonte-Savoia	2 × 600	320	190	Alstom	HVDC MaxSine
Dalian City Feed	1000	320	43	C-EPRI	HVDC Flexible
Xiamen Island HVDC	1000	320	10.7	C-EPRI	HVDC Flexible

and *HVDC MaxSine* to reach an operating voltage of ±320 kV [55]. These technologies are available with HB-SM, FB-SM, and hybrid SMs. The first MMC-based HVDC system is available with HB-SM with a power capacity of 400 MW at ±200 kV voltage. This system was installed in San Francisco's *Trans Bay* project by *Siemens* in 2010 [56]. Similarly, the *HVDC MaxSine* technology is available with FB-SM along with IGBT devices that are connected in series with them to form an MMC arm.

5.7.2 Offshore wind farms

Wind farms (WFs) are widely located offshore due to the high-energy harvest from the wind. These WFs are connected to the grid for subsequent distribution and consumption of generated power. The MMC-based HVDC transmission system is a suitable candidate to transmit the bulk power from the WFs to the grid compared with a line-commutated converter-based HVDC transmission system [48]. The MMC-based HVDC system provides a black start operation and eliminates the external harmonic filters to improve the grid power factor. In addition, the MMC-based HVDC system has a compact structure where it is possible to control both active and reactive powers. The circuit configuration of the MMC-based HVDC transmission system connected to an offshore WF through a subsea cable is shown in Fig. 5.15. The WF consists of a WF substation with a transformer, which will step up the WF voltage so that the subsea cable and converter losses can be reduced [57]. There is a wide range of projects based on MMC-based HVDC for offshore WFs with a maximum power capacity of 900 MW and voltage ±320 kV. Most of them are installed in the European region (see Table 5.7).

5.7.3 Medium-voltage motor drives

Conventional multilevel converters are widely used in motor drive applications which can reach an operating voltage up to 13.8 kV. These converters require a transformer, which is bulky and costly. On the other hand, the transformerless operation is possible with an MMC and is suitable for motor drive applications. Furthermore, a

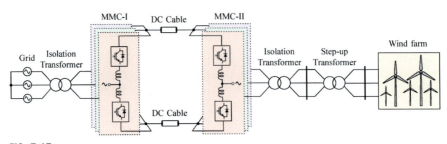

FIG. 5.15

MMC-HVDC with submarine cable for an offshore WF.

Table 5.7 Market overview of MMC-HVDC-based offshore WFs [11,18,54].

Project name	Power (MW)	Voltage (±kV)	DC cable (km)	Company	HVDC Tech.
Borwin-2	800	300	200	Siemens	HVDC Plus
Sylwin-1	864	320	205	Siemens	HVDC Plus
Dolwin-2	900	320	135	ABB	HVDC Light Gen. 4
Dolwin-3	900	320	160	Alstom	HVDC MaxSine
Nanhui	18	30	8.4	C-EPRI	HVDC Flexible

wide range of operating voltages can be achieved with an MMC by varying the number of SMs in each arm [58, 59]. In addition, the MMC can operate with standard motor drives without any output filters and offer a simple construction compared with conventional multilevel converters.

Motor drives are designed to achieve a four-quadrant operation including regenerative operation so that energy saving can be achieved. The typical configuration of the MMC-based four-quadrant motor drive is shown in Fig. 5.16. In this configuration, the MMC-I is connected to the medium-voltage grid, which is controlled with the VOC method to achieve control objectives such as DC-bus voltage control and grid power factor correction. The MMC-II is connected to the motor and it is controlled with either the FOC or direct torque control (DTC) method to control the motor speed and torque. Other possible solutions are replacing the grid-side MMC-I with a multi-pulse transformer (12-pulse or 18-pulse) and diode bridge rectifier, which simplifies the control complexity of the system. The multi-pulse transformer eliminates the dominant harmonic components from the grid-current and reduces the grid current harmonic distortion. Furthermore, the multi-pulse

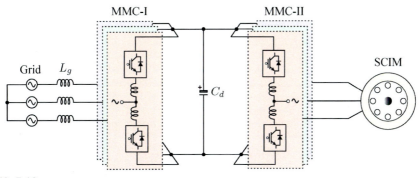

FIG. 5.16

MMC-based motor drive with active rectifier.

Table 5.8 Market overview of MMC-based motor drive systems [11, 18, 54].

Project name	Power (MVA)	Voltage (kV)	Developed by
Sinamics SM120 MV Drive	6–13.7	3.3–7.2	Siemens
M2L 3000 Series MV Drive	0.224–7.466	2.3–6.6	Benshaw

transformer blocks the common-mode current entering into the system such that the bearing failures and shaft torque pulsations can be prevented. Table 5.8 presents the commercial MMC-based motor drive systems, which are available with a maximum voltage of 7.2 kV and a power rating of 13.7 MVA by *Siemens* [60].

5.7.4 Power quality improvement

The distorted and unbalanced loads lead to power quality problems including voltage sag, voltage swell, reactive power compensation, harmonic distortion, and load unbalance. These problems at distribution and transmission levels are addressed with various power quality improvement devices such as a static VAr compensator (SVC), static synchronous compensator (STATCOM), dynamic voltage restorer (DVR), and unified power quality compensator (UPQC). These power quality devices mainly use a power converter as a building block. The conventional multilevel converters are widely used in power quality improvement, and these converters are connected to the medium-voltage grids through a transformer.

On the other hand, the MMC can reach a wide range of operating voltages by varying SMs in each arm and allows direct connection (without transformer) to the medium-voltage grid. The configuration of the MMC-based STATCOM is shown in Fig. 5.17. The main function of STATCOM is to compensate the load reactive power and improve the grid power factor [16]. There are several commercial MMC-based power quality improvement devices available in the market (refer to Table 5.9). In these applications, the MMC is designed with either an HB-SM or FB-SM. The first MMC-based STATCOM was developed by *Siemens* to handle a voltage of 33 kV without a transformer and 220 kV with a transformer [16]. These products were installed in various industries including offshore WFs, substations, and arc furnaces to supply the reactive power.

5.8 Conclusions

In this chapter, a comprehensive study on the operation, control, and application of an MMC is presented. An MMC uses an SM as a building block, which can be configured with various power electronic circuits. The operation and features of the most widely used SMs including HB-SM, FB-SM, FC-SM, and CHB-SM are discussed. However, the operation and performance of an MMC are analyzed with an HB-SM due to its simple structure, ease of control, and high efficiency. In addition, a

5.8 Conclusions

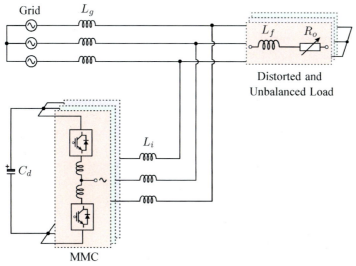

FIG. 5.17

Configuration of MMC-based STATCOM.

Table 5.9 Market overview of MMC-based STATCOM [11, 18, 54].

Project name	Power (±MVAr)	Voltage (kV)	Company	STATCOM Tech.
Kikiwa SVC Plus	2 × 35	220/11	Siemens	SVC Plus M
Mocuba Substation	35	33	Siemens	SVC Plus M
Inter Island Link pole 3	50	220	Siemens	SVC Plus C
Rio Branco SVC Plus	55	230	Siemens	SVC Plus C
Offshore Greater Gabbard	50	13.9	Siemens	SVC Plus L

PWM-based classical control method is presented to achieve various control objectives including SM capacitor voltage control, output current control, and circulating current reduction. An MMC requires a PWM scheme to generate the gating signals to the SMs. These PWM schemes are categorized into high, medium, and fundamental switching frequency PWM schemes, and their features and applications are briefly discussed. Among them, the staircase PWM is employed in an MMC-based HVDC application, whereas the phase-shifted carrier PWM scheme is used in an MMC-based motor drive application. Also, the control structure and implementation of SM capacitor voltage control, output current control, and circulating current control are presented. The SM capacitor voltage consists of a leg voltage control and voltage balancing strategy. The leg voltage control maintains the average leg voltage at its nominal value, whereas the voltage balancing strategy equally distributes the net DC-bus voltage between the SMs within the arm. The output current and circulating current control are implemented in the dq-frame, where the output current control ensures the perfect tracking of the reference and actual output currents so that the load performance will be improved significantly. The circulating current control is designed to minimize the second-harmonic component in the circulating current such that the magnitude of arm current will be kept within the rated value. The reduction of circulating current improves the converter efficiency and reliability. The analysis and results presented in this chapter prove the effectiveness of the PWM-based classical control method for an MMC. Finally, the commercial applications of an MMC are briefly summarized, which proves the importance and potentiality of an MMC for high-power industrial applications.

References

[1] C. Wang, Y. Li, Analysis and calculation of zero-sequence voltage considering neutral-point potential balancing in three-level npc converters, IEEE Trans. Ind. Electron. 57 (7) (2010) 2262–2271.

[2] L. He, C. Cheng, A flying-capacitor-clamped five-level inverter based on bridge modular switched-capacitor topology, IEEE Trans. Ind. Electron. 63 (12) (2016) 7814–7822.

[3] H. Salimian, H. Iman-Eini, Fault-tolerant operation of three-phase cascaded h-bridge converters using an auxiliary module, IEEE Trans. Ind. Electron. 64 (2) (2017) 1018–1027.

[4] E. Burguete, J. Lopez, M. Zabaleta, A new five-level active neutral-point-clamped converter with reduced overvoltages, IEEE Trans. Ind. Electron. 63 (11) (2016) 7175–7183.

[5] U.M. Choi, K.B. Lee, F. Blaabjerg, Diagnosis and tolerant strategy of an open-switch fault for t-type three-level inverter systems, IEEE Trans. Ind. Appl. 50 (1) (2014) 495–508.

[6] A. Dekka, M. Narimani, Capacitor voltage balancing and current control of a five-level nested neutral-point-clamped converter, IEEE Trans. Power Electron 33 (12) (2018) 10169–10177.

[7] D. Floricau, F. Richardeau, New multilevel converters based on stacked commutation cells with shared power devices, IEEE Trans. Ind. Electron. 58 (10) (2011) 4675–4682.

[8] H. Abu-Rub, J. Holtz, J. Rodriguez, G. Baoming, Medium-voltage multilevel converters: state of the art, challenges, and requirements in industrial applications, IEEE Trans. Ind. Electron. 57 (8) (2010) 2581–2596.

[9] S. Kouro, J. Rodriguez, B. Wu, S. Bernet, M. Perez, Powering the future of industry: high-power adjustable speed drive topologies, IEEE Ind. Appl. Mag. 18 (4) (2012) 26–39.

[10] A. Lesnicar and R. Marquardt, "An innovative modular multilevel converter topology suitable for a wide power range," in *Power Tech Conference Proceedings, 2003 IEEE Bologna*, vol. 3, 2003, pp. 23–26.

[11] M. Perez, S. Bernet, J. Rodriguez, S. Kouro, R. Lizana, Circuit topologies, modeling, control schemes, and applications of modular multilevel converters, IEEE Trans. Power Electron. 30 (1) (2015) 4–17.
[12] S. Debnath, J. Qin, B. Bahrani, M. Saeedifard, P. Barbosa, Operation, control, and applications of the modular multilevel converter: a review, IEEE Trans. Power Electron. 30 (1) (2015) 37–53.
[13] S. Teeuwsen, Modeling the trans bay cable project as voltage-sourced converter with modular multilevel converter design, in: *2011 IEEE Power and Energy Society General Meeting*, July, 2011, pp. 1–8.
[14] SIEMENS, "The smart way hvdc plus-one step ahead." [Online]. 2013 Available: https://www.energy.siemens.com/us/en/power-transmission/hvdc/hvdcplus/.
[15] BENSHAW, "M2l 3000 series medium voltage motor drive." [Online], 2013. Available: http://www.benshaw.com/uploadedFiles/Literature/Benshaw\M2L\MVFD\2.3-6.6kV.pdf.
[16] SIEMENS, "The efficient way." [Online], 2013. Available: http://www.energy.siemens.com/ru/pool/hq/power-transmission/FACTS/SVC\PLUS\ The\%20efficient\%20Way.pdf.
[17] A. Nami, J. Liang, F. Dijkhuizen, G. Demetriades, Modular multilevel converters for hvdc applications: review on converter cells and functionalities, IEEE Trans. Power Electron. 30 (1) (2015) 18–36.
[18] A. Dekka, B. Wu, R.L. Fuentes, M. Perez, N.R. Zargari, Evolution of topologies, modeling, control schemes, and applications of modular multilevel converters, IEEE Trans. Emerg. Sel. Topics Power Electron. 5 (4) (2017) 1631–1656.
[19] M.A. Perez, F.R. Lizana, J. Rodrıguez, Decoupled current control of modular multilevel converter for hvdc applications, in: *2012 IEEE International Symposium on Industrial Electronics*, May, 2012, pp. 1979–1984.
[20] M. Hagiwara, K. Nishimura, H. Akagi, A medium-voltage motor drive with a modular multilevel pwm inverter, IEEE Trans. Power Electron. 25 (7) (2010) 1786–1799.
[21] E. Solas, G. Abad, J. Barrena, S. Aurtenetxea, A. Carcar, L. Zajac, Modular multilevel converter with different submodule concepts—part i: capacitor voltage balancing method, IEEE Trans. Ind. Electron. 60 (10) (2013) 4525–4535.
[22] M. Hagiwara, H. Akagi, Control and experiment of pulsewidth-modulated modular multilevel converters, IEEE Trans. Power Electron. 24 (7) (2009) 1737–1746.
[23] X. She, A. Huang, X. Ni, R. Burgos, Ac circulating currents suppression in modular multilevel converter, in: *IECON 2012 - 38th Annual Conference on IEEE Industrial Electronics Society*, Oct, 2012, pp. 191–196.
[24] R. Darus, J. Pou, G. Konstantinou, S. Ceballos, V.G. Agelidis, Controllers for eliminating the ac components in the circulating current of modular multilevel converters, IET Power Electron. 9 (1) (2016) 1–8.
[25] Y. Li, F. Wang, Arm inductance selection principle for modular multilevel converters with circulating current suppressing control, in: *Applied Power Electronics Conference and Exposition (APEC), 2013 Twenty-Eighth Annual IEEE*, Mar 2013, 2013, pp. 1321–1325.
[26] B. Bahrani, S. Debnath, M. Saeedifard, Circulating current suppression of the modular multilevel converter in a double-frequency rotating reference frame, IEEE Trans. Power Electron. 31 (1) (2016) 783–792.
[27] Z. Li, P. Wang, Z. Chu, H. Zhu, Y. Luo, Y. Li, An inner current suppressing method for modular multilevel converters, IEEE Power Electron. Lett. 28 (11) (2013) 4873–4879.

[28] P. Meshram, V. Borghate, A simplified nearest level control (nlc) voltage balancing method for modular multilevel converter (mmc), IEEE Trans. Power Electron. 30 (1) (2015) 450–462.

[29] R. Darus, G. Konstantinou, J. Pou, S. Ceballos, V.G. Agelidis, Comparison of phase-shifted and level-shifted pwm in the modular multilevel converter, in: *2014 International Power Electronics Conference (IPEC-Hiroshima 2014 - ECCE ASIA)*, May, 2014, pp. 3764–3770.

[30] G. Konstantinou, J. Zhang, S. Ceballos, J. Pou, V.G. Agelidis, Comparison and evaluation of sub-module configurations in modular multilevel converters, in: *2015 IEEE 11th International Conference on Power Electronics and Drive Systems*, June, 2015, pp. 958–963.

[31] J. Kolb, F. Kammerer, M. Gommeringer, M. Braun, Cascaded control system of the modular multilevel converter for feeding variable-speed drives, IEEE Trans. Power Electron. 30 (1) (2015) 349–357.

[32] A. Dekka, B. Wu, N.R. Zargari, R.L. Fuentes, Dynamic voltage balancing algorithm for modular multilevel converter: a unique solution, IEEE Trans. Power Electron. 31 (2) (2016) 952–963.

[33] E. Solas, G. Abad, J.A. Barrena, A. Carear, S. Aurtenetxea, Modelling, simulation and control of modular multilevel converter, in: *Proceedings of 14th International Power Electronics and Motion Control Conference EPE-PEMC 2010, Sept*, 2010. pp. T2–90-T2–96.

[34] S. Du, J. Liu, T. Liu, Modulation and closed-loop-based dc capacitor voltage control for mmc with fundamental switching frequency, IEEE Trans. Power Electron. 30 (1) (2015) 327–338.

[35] A. Dekka, B. Wu, N.R. Zargari, Dynamic voltage balancing algorithm for modular multilevel converter with three-level flying capacitor submodules, in: *2014 International Power Electronics Conference (IPEC-Hiroshima 2014 - ECCE-ASIA)*, May, 2014, pp. 3468–3475.

[36] A. Hassanpoor, S. Norrga, H.P. Nee, L. Angquist, Evaluation of different carrier-based pwm methods for modular multilevel converters for hvdc application, in: *IECON 2012 - 38th Annual Conference on IEEE Industrial Electronics Society, Oct*, 2012, pp. 388–393.

[37] Q. Tu, Z. Xu, L. Xu, Reduced switching-frequency modulation and circulating current suppression for modular multilevel converters, IEEE Trans. Power Del. 26 (3) (2011) 2009–2017.

[38] Z. Li, P. Wang, H. Zhu, Z. Chu, Y. Li, An improved pulse width modulation method for chopper-cell-based modular multilevel converters, IEEE Trans. Power Electron. 27 (8) (2012) 3472–3481.

[39] D. Siemaszko, A. Antonopoulos, K. Ilves, M. Vasiladiotis, L. Angquist, H.-P. Nee, Evaluation of control and modulation methods for modular multilevel converters, in: *Power Electronics Conference (IPEC), 2010 International*, Jun, 2010, pp. 746–753.

[40] A. Dekka, B. Wu, N. Zargari, A novel modulation scheme and voltage balancing algorithm for modular multilevel converter, in: *2015 IEEE Applied Power Electronics Conference and Exposition (APEC), Charlotte, NC*, 2015, pp. 1227–1233.

[41] A. Dekka, B. Wu, N.R. Zargari, R.L. Fuentes, A space-vector pwm-based voltage-balancing approach with reduced current sensors for modular multilevel converter, IEEE Trans. Ind. Electron. 63 (5) (2016) 2734–2745.

[42] A. Dekka, B. Wu, N.R. Zargari, A novel modulation scheme and voltage balancing algorithm for modular multilevel converter, IEEE Trans. Ind. Appl. 52 (1) (2016) 432–443.

[43] P. Hu, D. Jiang, A level-increased nearest level modulation method for modular multi-level converters, IEEE Trans. Power Electron. 30 (4) (2015) 1836–1842.
[44] G. Konstantinou, M. Ciobotaru, V. Agelidis, Selective harmonic elimination pulse-width modulation of modular multilevel converters, IET Power Electron. 6 (1) (2013) 96–107.
[45] M. Saeedifard, R. Iravani, Dynamic performance of a modular multilevel back-to-back hvdc system, IEEE Trans. Power Del. 25 (4) (2010) 2903–2912.
[46] J.J. Jung, S. Cui, J.H. Lee, S.K. Sul, A new topology of multilevel vsc converter for a hybrid hvdc transmission system, IEEE Trans. Power Electron. 32 (6) (2017) 4199–4209.
[47] G. Liu, F. Xu, Z. Xu, Z. Zhang, G. Tang, Assembly hvdc breaker for hvdc grids with modular multilevel converters, IEEE Trans. Power Electron. 32 (2) (2017) 931–941.
[48] H. Liu, K. Ma, Z. Qin, P.C. Loh, F. Blaabjerg, Lifetime estimation of mmc for offshore wind power hvdc application, IEEE Trans. Emerg. Sel. Topics Power Electron. 4 (2) (2016) 504–511.
[49] U.N. Gnanarathna, S.K. Chaudhary, A.M. Gole, R. Teodorescu, Modular multi-level converter based hvdc system for grid connection of offshore wind power plant, in: *9th IET International Conference on AC and DC Power Transmission (ACDC 2010)*, Oct, 2010, pp. 1–5.
[50] Y. Long, X. Xiao, Y. Xu, B. Yu, Y. Xu, J. Hao, Mmc-upqc: Application of modular multilevel converter on unified power quality conditioner, in: *2013 IEEE Power Energy Society General Meeting*, July, 2013, pp. 1–5.
[51] R.E. Torres-Olguin, M. Molinas, T. Undeland, Offshore wind farm grid integration by vsc technology with lcc-based hvdc transmission, IEEE Trans. Sustainable Energy 3 (4) (2012) 899–907.
[52] N. Flourentzou, V.G. Agelidis, G.D. Demetriades, Vsc-based hvdc power transmission systems: an overview, IEEE Trans. Power Electron. 24 (3) (2009) 592–602.
[53] A. Beddard, M. Barnes, Modelling of mmc-hvdc systems–an overview, Energy Procedia 80 (2015) 201–212.
[54] S. Du, A. Dekka, B. Wu, N. Zargari, Modular Multilevel Converters: Analysis, Control, and Applications, Wiley-IEEE Press, Hoboken, NJ, 2017.
[55] H.J. Knaak, Modular multilevel converters and hvdc/facts: A success story, in: *Power Electronics and Applications (EPE 2011), Proceedings of the 2011-14th European Conference on*, Aug, 2011, pp. 1–6.
[56] K. Friedrich, Modern hvdc plus application of vsc in modular multilevel converter topology, in: *2010 IEEE International Symposium on Industrial Electronics*, July, 2010, pp. 3807–3810.
[57] J. Glasdam, J. Hjerrild, L.H. Kocewiak, C.L. Bak, Review on multi-level voltage source converter based hvdc technologies for grid connection of large offshore wind farms, in: *Power System Technology (POWERCON), 2012 IEEE International Conference on*, Oct, 2012, pp. 1–6.
[58] J.-J. Jung, H.-J. Lee, S.-K. Sul, Control strategy for improved dynamic performance of variable-speed drives with modular multilevel converter, IEEE Trans. Emerg. Sel. Topics Power Electron. 3 (2) (2015) 371–380.
[59] S. Du, B. Wu, K. Tian, N.R. Zargari, Z. Cheng, An active cross-connected modular multilevel converter (ac-mmc) for a medium-voltage motor drive, IEEE Trans. Ind. Electron. 63 (8) (2016) 4707–4717.
[60] SIEMENS, "Sinamics sm120." [Online], 2013. Available: http://www.industry.siemens.com/drives/global/en/converter/mv-drives/sinamics-sm120-cm/.

CHAPTER 6

Asymmetrical multilevel inverter topologies

Ilhami Colak[a], Ersan Kabalcı[b], and Gokhan Keven[c]

[a]Department of Electrical and Electronics Engineering, Faculty of Engineering and Architecture, Nisantasi University, Istanbul, Turkey [b]Department of Electrical and Electronics Engineering, Faculty of Engineering and Architecture, Nevsehir Haci Bektas Veli University, Nevsehir, Turkey [c]Department of Electronics and Automation, Vocational High School, Nevsehir Haci Bektas Veli University, Hacibektas, Nevsehir, Turkey

6.1 Introduction

The use of renewable energy sources (RESs) is increasing, due to the need to eliminate the harmful effects and environmental pollution caused by fossil fuel use. The decreasing reserves and increasing cost of fossil-based fuels are also other reasons forcing the search for alternative fuels. The generated electrical energy obtained from RESs usually needs to be converted and regulated at certain voltage waveforms and levels due to the intermittent structure of the sources. Efficiency and sustainability are major issues in DC/AC conversion (inverter) systems. Inverter topologies are classified as traditional two-level and multilevel inverter structures (MLI). Two-level inverters are not practicable in medium voltage because of blockage voltage limitation and voltage stress on semiconductors. On the other hand, widely known traditional MLI topologies are classified as diode clamped MLI, flying capacitor MLI, and cascaded MLI, where diode clamped and flying capacitor topologies need a single DC source. The major disadvantage of cascaded MLI is its separate DC source requirement for generating staircase output. In spite of this, cascaded MLI topology has many advantages, such as less total harmonic distortion (THD), modular structure, increased voltage levels by series connection of modules, and fewer component numbers for any output level [1–3].

Cascaded MLI topologies are categorized based on the ratio of separated DC sources, where symmetrical topologies are fed by identical source levels while the asymmetrical ones are based on multiple ratios of based DC source. The multiple feeding source ratios of asymmetrical MLI (A-MLI) provide an increased number of levels with the same number of components and same configuration of symmetrical MLI. A-MLI topologies are developed for reducing the size and cost of MLI with decreased THD ratios [4, 5].

Many different A-MLI topologies have been proposed in order to achieve different voltage levels. Output voltage levels are produced by using a cascaded connection of base units of an A-MLI structure. These topologies can be classified into two main categories, as shown in Fig. 6.1, where the first category uses a polarity changer in the configuration. Two sections are available in this topology: the level generation stage and polarity generation stage. The generation stage produces zero and positive levels while the polarity generation stage is in charge of transferring the positive and negative voltage levels based on the H-bridge structure. The topologies without a polarity changer use H-bridge or reduced number component topologies in the configuration. Reduced number component topologies are another category for achieving increased voltage levels by using fewer components in the structure, such as switching components, power diodes, driver circuits, and capacitors [5].

While the A-MLIs are classified as binary or trinary due to the ratios of DC sources at the input, some A-MLI topologies using an irregular ratio of DC sources have been proposed in the literature. These inverters which use an irregular ratio of DC sources, can be called disordered or irregular topologies due to unusual source levels. Fig. 6.2 depicts output voltage levels of symmetric MLI, binary type A-MLI, and trinary type A-MLI, respectively. The inverter configurations are used as regular cascaded H-bridge for generating these output waveforms. All these topologies are fed by two DC sources at the input and eight switching devices in two H-bridges. Once the symmetric topology is fed by two separate DC sources as $V_1 = V_2 = V_{DC}$, the output voltage levels of the inverter are obtained as $+2V_{DC}, +V_{DC}, 0, -V_{DC}, -2V_{DC}$.

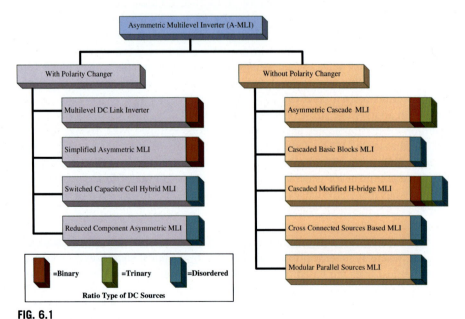

FIG. 6.1

A-MLI topologies.

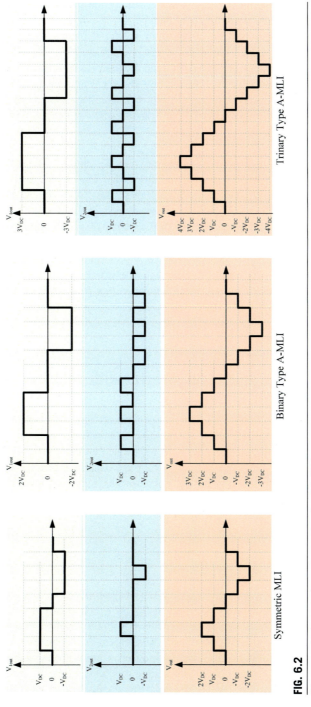

FIG. 6.2

Binary and trinary voltage output levels of A-MLI.

Despite symmetrical topologies, an A-MLI topology fed in binary DC source level generates 7-level output, while the trinary type A-MLI generates 9-level output. If the DC sources are defined as $V_1 = V_{DC}$ and $V_2 = 2V_{DC}$ in binary type A-MLI, the output values of the inverter are obtained at $+3V_{DC}$, $+2V_{DC}$, $+V_{DC}$, 0, $-V_{DC}$, $-2V_{DC}$, $-3V_{DC}$. If the DC sources are $V_1 = V_{DC}$ and $V_2 = 3V_{DC}$ in a trinary type A-MLI, the output voltage levels of the inverter will be $+4V_{DC}$, $+3V_{DC}$, $+2V_{DC}$, $+V_{DC}$, 0, $-V_{DC}$, $-2V_{DC}$, $-3V_{DC}$, $-4V_{DC}$, as illustrated in Fig. 6.2. The output levels of the inverters show that A-MLI topologies achieve more levels when compared with symmetric MLI [5, 6].

The polarity generation part is an H-bridge module, as depicted in Fig. 6.3A, where four switches exist in a single module (S_1, S_2, S_3, and S_4). There are two states for zero voltage level as shown in Fig. 6.3B and C: S_1–S_3 are in the "ON" and S_2–S_4 are in the "OFF" position or S_2–S_4 are in the "ON" and S_1–S_3 are in the "OFF" position. Positive voltage level generation is produced by S_1–S_4 in "ON" and S_2–S_3 in "OFF" position, as shown in Fig. 6.3D. Negative voltage level generation is generated by S_2–S_3 in "ON" and S_1–S_4 in "OFF" position, as depicted in Fig. 6.3E. Additionally, the polarity generation part is switched at the fundamental frequency while the level generation part is switched at high frequency [5].

The structure of A-MLI topologies, their operating principles, component numbers, calculation of output levels, switching sequences, and comparison of topologies based on these values are described in the following subsections.

6.2 Asymmetric multilevel inverter with polarity generation part

When an A-MLI is composed with a level generation part, the inverter must have a polarity generation part for obtaining positive and negative periods. The polarity generation part is an H-bridge module, as mentioned earlier. Many different topologies exist for the level generation part, which are usually connected in series. The number of units must be increased for achieving an increased level at the output. However, an increased number of units increases the separate DC source requirement for every level generation unit. A block diagram of an A-MLI using a polarity generation part is shown in Fig. 6.4.

6.2.1 Multilevel DC link inverter

A multilevel DC link inverter (ML-DCLI) has been proposed by Gui-Jia Su [7]. The topology is based on half-bridge modules in the level generation part and an H-bridge in the polarity generation part, as seen in Fig. 6.5. In the level generation part, half-bridge modules are supplied by separated DC sources. Two complementary switches in one half-bridge module control the DC source. Output levels of the inverter are achieved with series connection of half-bridge modules, where the level generation part produces only positive levels. The polarity generation part must be operated to

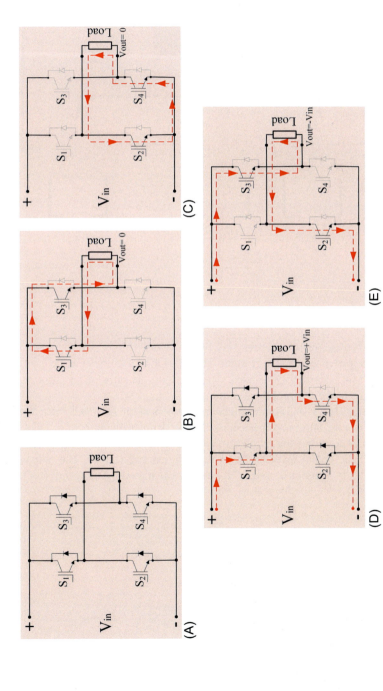

FIG. 6.3

Switching modes of the polarity generation part: (A) H-bridge, (B) Zero state1, (C) Zero state2, (D) Positive state, (E) Negative state.

CHAPTER 6 Asymmetrical multilevel inverter topologies

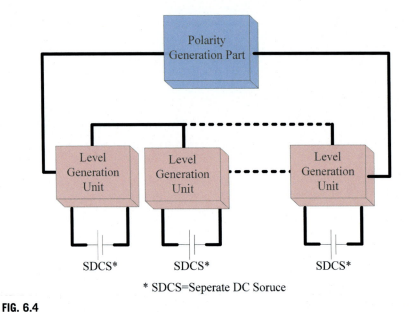

FIG. 6.4

General diagram for A-MLI using polarity generation part.

produce negative levels at the output. The ML-DCLI topology can be used for symmetrical and asymmetrical modes, depending on the ratio of DC sources that are binary type in the asymmetrical mode [5, 7].

Equations (6.1)–(6.3) represent the parameters of the ML-DCLI in asymmetrical mode. When the number of base units is n, N_S denotes the number of DC sources in Eq. (6.1), N_{SW} shows the number of switching components in Eq. (6.2), and N_{L-A} calculates the output level of the inverter by using Eq. (6.3). The output level of the inverter in symmetric mode (N_L) can be calculated by using Eq. (6.4) [5, 7].

$$N_S = n \tag{6.1}$$

$$N_{SW} = 2n + 4 \tag{6.2}$$

$$N_{L-A} = 2^{(n+1)} - 1 \tag{6.3}$$

$$N_{L-S} = 2n + 1 \tag{6.4}$$

Fig. 6.6 illustrates the advantages of A-MLI, while comparing symmetric and asymmetric modes for switching device number and output levels. The same number of switching devices is used for symmetric and asymmetric modes as calculated from the equations given earlier. While the number of DC sources is increased, the switching device number is also increased in the same way. The output level of the

6.2 Asymmetric multilevel inverter with polarity generation part

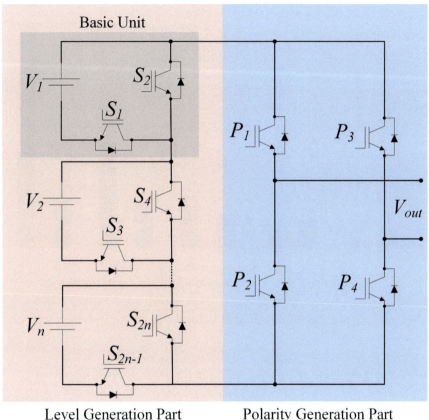

FIG. 6.5

General structure of ML-DCLI.

symmetric topology increases depending on DC sources. However, the output level increases drastically in asymmetric mode when compared to symmetric mode.

Table 6.1 shows the switching states and output voltage levels for the level generation part in a positive sequence for a 31-level ML-DCLI that can be achieved with four separate DC sources and eight switching devices in the level generation part. Total switch number becomes 12 with additional H-bridge switches in the polarity generation part. The DC sources are fixed in the binary type as $V_1 = V_{DC}$, $V_2 = 2V_{DC}$, $V_3 = 4V_{DC}$ and $V_4 = 8V_{DC}$ voltage ratios. The maximum magnitude is the sum of DC sources. The positive levels are transferred via the polarity generation part as positive or negative levels.

The H-bridge switches must be turned "OFF" while the level generation part switches are turned "ON" in both operations as zero voltage or zero current for inductive load in the ML-DCLI. This approach has been recommended for reducing the

188 CHAPTER 6 Asymmetrical multilevel inverter topologies

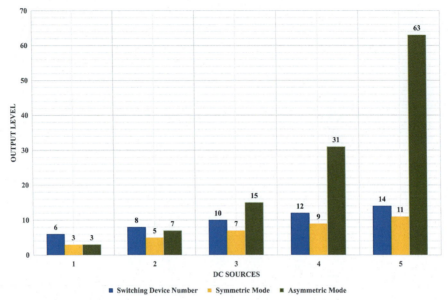

FIG. 6.6

Output levels and switching device number of ML-DCLI in symmetric and asymmetric mode.

Table 6.1 Switching sequences of positive levels for 31-level ML-DCLI.

		Level generation part							
V_{out}	Level	S_1	S_2	S_3	S_4	S_5	S_6	S_7	S_8
0	0	0	1	0	1	0	1	0	1
V	V_1	1	0	0	1	0	1	0	1
2V	V_2	0	1	1	0	0	1	0	1
3V	$V_1 + V_2$	1	0	1	0	0	1	0	1
4V	V_3	0	1	0	1	1	0	0	1
5V	$V_3 + i_1$	1	0	0	1	1	0	0	1
6V	$V_3 + V_2$	0	1	1	0	1	0	0	1
7V	$V_3 + V_2 + V_1$	1	0	1	0	1	0	0	1
8V	V_4	0	1	0	1	0	1	1	0
9V	$V_4 + V_1$	1	0	0	1	0	1	1	0
10V	$V_4 + V_2$	0	1	1	0	0	1	1	0
11V	$V_4 + V_2 + V_1$	1	0	1	0	0	1	1	0
12V	$V_4 + V_3$	0	1	0	1	1	0	1	0
13V	$V_4 + V_3 + V_1$	1	0	0	1	1	0	1	0
14V	$V_4 + V_3 + V_2$	0	1	1	0	1	0	1	0
15V	$V_4 + V_3 + V_2 + V_1$	1	0	1	0	1	0	1	0

switching losses in the H-bridge. The ML-DCLI is switched in two different frequencies, where the level generation part is operated at high frequency while the polarity generation part is switched at the fundamental frequency. Therefore the H-bridge switching losses are reduced in this way. The benefits of this topology are the possibility of equal load sharing, smaller conduction losses, and simplicity in structure. The advantages of the ML-DCLI can be noted as decreasing the requirement for an additional H-bridge for negative levels and the number of redundant switching states [2, 5, 7].

6.2.2 Simplified asymmetric multilevel inverter

Researchers have proposed an alternative A-MLI topology, which is highlighted in Fig. 6.7. This simplified asymmetric multilevel inverter (SA-MLI) can be used in both symmetric and asymmetric modes. The level generation part is obtained by serial connection of basic units. One basic unit comprises a DC source, a switch, and a diode. The level generation part is used for achieving the positive levels, while the polarity generation part defines the positive or negative periods. The DC sources can be used as binary type when the SA-MLI topology is used in asymmetrical mode. Eqs (6.5) and (6.6) represent the parameters of the SA-MLI in asymmetrical mode. The advantages of the SA-MLI inverter are a minimum number of switching components, semiconductor gate drivers, and decreased inverter cost. Two drawbacks exist in the SA-MLI, which are the extra high voltage requirement for the H-bridge and decreased modularity [8–13].

The output voltage level of the inverter, N_L, is obtained by using Eq. (6.5), where the number of basic units and diodes are denoted by n. The N_{SW} represents the total number of switching components in the inverter topology, as depicted in Eq. (6.6) [8–13].

$$N_L = 2^{(n-1)} - 1 \tag{6.5}$$

$$N_{SW} = n + 4 \tag{6.6}$$

Fig. 6.8 shows the variation of output voltage levels obtained by using certain numbers of switching devices and DC sources in symmetric and asymmetric modes. Each column has been obtained by defining an identical number of switching devices and DC sources in symmetric and asymmetric modes. The increment of DC source depends on increasing the basic unit number that causes an increment on number of switching devices. The output level is remarkably increased in the asymmetric operation mode compared to the symmetric mode. When the number of separate DC sources used in the topology is selected as 5, the output level of the asymmetric mode is increased to 63, while it was 11 in the symmetric operation mode. This result proves that the A-MLI topology is more efficient than symmetric topology. Also, SA-MLI topologies have some benefits, including reduction of switching components, simple control strategies, and design flexibility [8–13].

190 CHAPTER 6 Asymmetrical multilevel inverter topologies

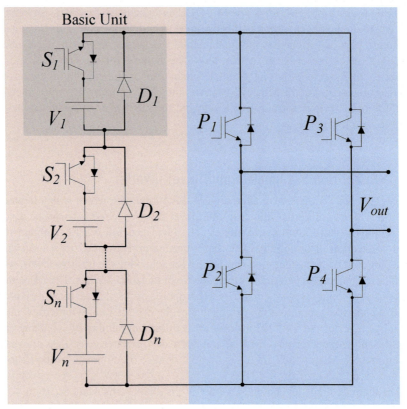

FIG. 6.7

General structure of SA-MLI.

Table 6.2 shows the switching states and output voltage levels for the level generation part in positive sequence for a 31-level SA-MLI that can be achieved with four separate DC sources supplying four switches and four diodes in the level generation part. In this topology, the total number of switches is eight, with adding H-bridge switches in the polarity generation part. DC sources are binary type and the values are $V_1 = V_{DC}$, $V_2 = 2V_{DC}$, $V_3 = 4V_{DC}$ and $V_4 = 8V_{DC}$ [13].

Doss et al. [9] performed a simulation of an SA-MLI inverter. The DC source ratios were set to binary type (1:2:4) and three sources were used at magnitudes of 15V, 30V, and 60V. The output voltage waveform and harmonic spectrum of the modeled study are given in Fig. 6.9. The SA-MLI topology is a 15-level for three separate DC sources in asymmetric mode and the maximum magnitude of the inverter is 105V, that is, the sum of the DC sources. The harmonic spectrum shows that the voltage THD of the inverter is 3.51% without a filter [9].

6.2 Asymmetric multilevel inverter with polarity generation part

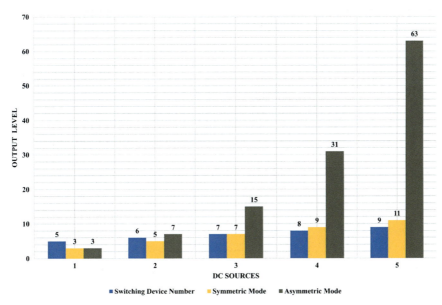

FIG. 6.8

Output levels and switching device number of SA-MLI in symmetric and asymmetric mode.

Table 6.2 Switching sequences of positive levels for 31-level SA-MLI.

		Level generation part			
V_{out}	Level	S_1	S_2	S_3	S_4
0	0	0	0	0	0
V	V_1	1	0	0	0
2V	V_2	0	1	0	0
3V	$V_1 + V_2$	1	1	0	0
4V	V_3	0	0	1	0
5V	$V_3 + V_1$	1	0	1	0
6V	$V_3 + V_2$	0	1	1	0
7V	$V_3 + V_2 + V_1$	1	1	1	0
8V	V_4	0	0	0	1
9V	$V_4 + V_1$	1	0	0	1
10V	$V_4 + V_2$	0	1	0	1
11V	$V_4 + V_2 + V_1$	1	1	0	1
12V	$V_4 + V_3$	0	0	1	1
13V	$V_4 + V_3 + V_1$	1	0	1	1
14V	$V_4 + V_3 + V_2$	0	1	1	1
15V	$V_4 + V_3 + V_2 + V_1$	1	1	1	1

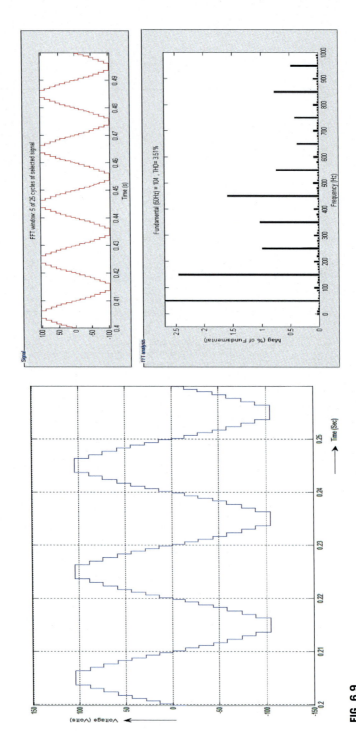

FIG. 6.9

Output voltage waveforms and switching device number of SA-MLI in symmetric and asymmetric mode [9].

6.2.3 Switched capacitor cell hybrid multilevel inverter

The switched capacitor cell hybrid MLI (SCH-MLI) is a combination of an H-bridge and a capacitor cell stage. This combination produces a basic unit and allows combining units by series connection for generating higher output levels. Two units of SCH-MLI topology are depicted in Fig. 6.10, where a capacitor cell has two switches and one diode for charge and discharge of the capacitor in each unit. When P_2 is turned to the "ON" position, the capacitor is charged to the DC source voltage. The capacitor charge voltage is added to the DC source when P_1 is turned to the "ON" position. The D_1 diode blocks the discharge of the capacitor through the DC sources when P_1 is turned on. This capacitor cell produces only positive levels and it operates as a level generation part. H-bridges of the SCH-MLI unit act for polarity generation. The SCH-MLI topology can be used for both symmetric and asymmetric mode. If the input DC source is $V_{DC} = V$ in one unit of the SCH-MLI, there are five levels at the output of the H-bridge that are $2V, V, 0, -V$ and $-2V$ [5, 14].

The SCH-MLI topology can be used at different ratios in each basic unit in asymmetric mode. However, some of the DC source algorithms are not able to achieve all levels. If there are "n" basic units, the number of DC sources required at different values and their magnitudes can be calculated by Eq. (6.7). The number of switching components is given in Eq. (6.8), where diode numbers (N_D) and capacitor numbers (N_C) can be calculated in the same way by Eq. (6.9). The total device number is determined by adding two devices (diode and capacitor) for each basic unit to the

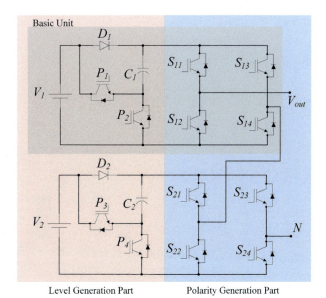

FIG. 6.10

General structure of SCH-MLI.

switching device number. Finally, output voltage level N_L is calculated using Eq. (6.10) [14].

$$V_{DC} = 5^{(n-1)} \cdot V_{dc} \quad (6.7)$$

$$N_{SW} = 6n \quad (6.8)$$

$$N_D = N_C = n \quad (6.9)$$

$$N_L = 5^n \quad (6.10)$$

Fig. 6.11 shows a parameter comparison for symmetric and asymmetric modes at different DC sources in an SCH-MLI. The output level of the inverter is incremented dramatically while DC sources are increased in the asymmetric mode when compared with the symmetric mode.

Table 6.3 illustrates the output levels of a single unit when the magnitude of the DC source is V. The SCH-MLI topology achieves five levels for a single basic unit with magnitudes of $2V, V, 0, -V$ and $-2V$. While the DC sources are at magnitudes of V and $5V$, the output voltage levels of series connected SCH-MLI units are shown in Table 6.4. The 25-level SCH-MLI can be achieved with two separate DC sources and there are four switches and two diodes in the level generation part. The number

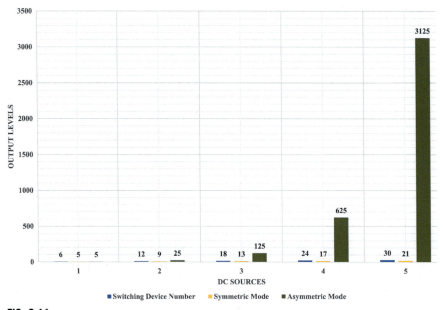

FIG. 6.11

Output levels and switching device number of the SCH-MLI in symmetric and asymmetric mode.

6.2 Asymmetric multilevel inverter with polarity generation part

Table 6.3 Output levels for one unit in SCH-MLI.

| V_{out} | Level | Level generation part || Polarity generation part ||||
		P_1	P_2	S_{11}	S_{12}	S_{13}	S_{14}
2V	V_2	1	0	1	0	0	1
V	V_1	0	1	1	0	0	1
0	0	0	1	0	1	0	1
−V	$-V_1$	0	1	0	1	1	0
−2V	$-V_2$	1	0	0	1	1	0

Table 6.4 Output levels for two series connected unit in SCH-MLI.

Level	Unit 1	Unit 2	Output value	Level	Unit 1	Unit 2	Output value
1	0	0	0	14	−V	0	−V
2	V	0	V	15	−2V	0	−2V
3	2V	0	2V	16	2V	−5	−3V
4	−2V	5	3V	17	V	−5	−4V
5	−V	5	4V	18	0	−5	−5V
6	0	5	5V	19	−V	−5	−6V
7	V	5	6V	20	−2V	−5	−7V
8	2V	5	7V	21	2V	−10	−8V
9	−2V	10	8V	22	V	−10	−9V
10	−V	10	9V	23	0	−10	−10V
11	0	10	10V	24	−V	−10	−11V
12	V	10	11V	25	−2V	−10	−12V
13	2V	10	12V				

of total switches is 12 with adding H-bridge switches in the polarity generation part. The main advantage of this topology is to enable an increment of DC source voltage to the double without any transformer requirement. The other advantage is that output levels can be obtained by adding and subtracting from each other. This operation generates more levels at the output as with the trinary type. The disadvantages of this topology are the capacitor ripple losses and conduction losses [5, 14].

6.2.4 Reduced component asymmetric multilevel inverter

Arif et al. [15] proposed an asymmetric topology called the reduced component asymmetric MLI (RCA-MLI) that uses level genesration and polarity generation parts together. The general structure of this topology is given in Fig. 6.12. The RCA-MLI can be extended for more levels by using additional basic units containing

196 CHAPTER 6 Asymmetrical multilevel inverter topologies

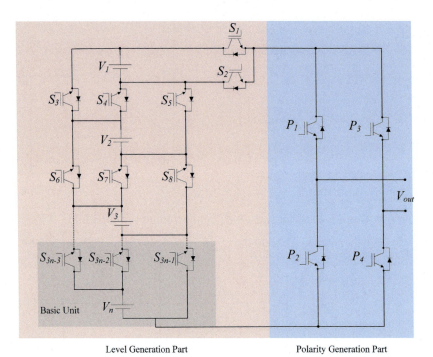

FIG. 6.12

General structure of RCA-MLI.

three switches and a single DC source. The DC source and two switches are permanent components in the level generation part. A disordered ratio is used in the DC supply section of the RCA-MLI topology [15–17].

The parameters of the RCA-MLI are calculated in the following equations. If the number of basic units is "n," the number of DC sources N_S is calculated as seen in Eq. (6.11). Two switches in the level generation part and four switches in the polarity generation part are constant in topology. Therefore, six switches are added to the basic units. The total switch number (N_{SW}) is calculated in Eq. 6.11 while the value of DC sources in one unit ($V_{DC,n}$) is determined according to the magnitude of the permanent DC source, as shown in Eq. (6.12). When two basic units are used in the RCA-MLI topology with a permanent source $V_1 = V$, the values of V_2, V_3 will be $3V$ and $8V$, respectively. The output level of the inverter is calculated by including the DC source value of the units to the permanent DC source. N_L is the output level of the inverter in Eq. (6.13). The switching states for a 9-level RCA-MLI topology are depicted in Table 6.5 [15–17].

$$N_S = n + 1 \tag{6.11}$$

$$N_{SW} = 3n + 6 \tag{6.12}$$

Table 6.5 Switching states for 9-level RCA-MLI topology.

	Level generation part						Polarity generation part				Output voltage
Level	S_1	S_2	S_3	S_4	S_5	S_6	P_1	P_2	P_3	P_4	
1	1	0	0	1	0	1	1	0	0	1	4V
2	0	1	0	1	0	1	1	0	0	1	3V
3	0	1	1	0	0	1	1	0	0	1	2V
4	1	0	0	0	1	0	1	0	0	1	1V
5	0	1	0	0	1	0	1	0	1	0	0
6	1	0	0	1	0	0	0	1	1	0	−1V
7	0	1	0	1	0	0	0	1	1	0	−2V
8	0	1	1	0	0	0	0	1	1	0	−3V
9	1	0	0	0	1	0	0	1	1	0	−4V

$$V_{DC,n} = \left(1 + V_n + \sum_{i=1}^{n} V_i\right) \times V_1 \tag{6.13}$$

$$N_L = 1 + \left\{2 \times \sum_{i=1}^{n} V_i\right\} \tag{6.14}$$

6.3 Asymmetric MLI topologies without polarity generation module

6.3.1 Asymmetric cascade multilevel inverter

Asymmetric cascade multilevel inverter (AC-MLI) topologies are based on a regular cascade multilevel topology, where the H-bridges are connected in series with a different ratio of DC sources. The AC-MLI topology shown in Fig. 6.13 can be used in symmetric or asymmetric mode, where an increased output level of the inverter is seen when compared with symmetrical mode, as mentioned before. The AC-MLI can be used either in binary or trinary ratio of DC sources. If there are "n" number of H-bridge modules used in the AC-MLI, N_S shows the number of DC sources in Eq. (6.15). Each H-bridge has four switches, so the total number of switching components (N_{SW}) is calculated as seen in Eq. (6.16) and the output voltage level is calculated based on the ratio of DC sources. The $N_{L\text{ (binary)}}$ is the output level of the inverter for binary ratio and $N_{L(\text{trinary})}$ is the output level of the inverter for the trinary ratio in Eqs (6.17) and (6.18), respectively [6, 18].

198 CHAPTER 6 Asymmetrical multilevel inverter topologies

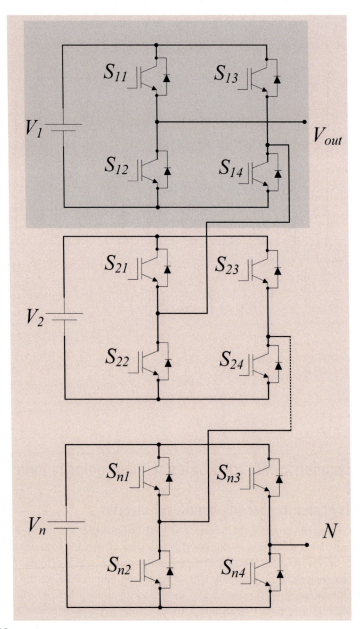

FIG. 6.13
General structure of AC-MLI.

6.3 Asymmetric MLI topologies without polarity generation module

Table 6.6 Switching states for seven-level AC-MLI topology.

V_{out}	Level	\multicolumn{4}{c	}{H-bridge unit-1}	\multicolumn{4}{c}{H-bridge unit-2}
		S_{11}	S_{12}	S_{13}	S_{14}	S_{21}	S_{22}	S_{23}	S_{24}
3V	$V_1 + V_2$	1	0	0	1	1	0	1	0
2V	V_2	0	1	0	1	1	0	1	0
V	V_1	1	0	0	1	0	1	0	1
0	0	0	0	0	0	0	0	0	0
−V	$-V_1$	0	1	1	0	0	1	0	1
−2V	$-V_2$	0	1	0	1	0	1	1	0
−3V	$-(V_1 + V_2)$	0	1	1	0	0	1	1	0

Table 6.6 shows the switching states and output voltage levels of two series connected H-bridge units for binary ratios of DC sources in AC-MLI. A 31-level AC-MLI can be achieved with 4 separate DC sources and 16 switches in 4 H-bridge units. When the DC sources are set at binary type, their values would be $V_1 = V$, $V_2 = 2V$, $V_3 = 4V$ and $V_4 = 8V$. If an AC-MLI is used in trinary ratio DC sources with 4 H-bridges, the output level of the inverter increases to 81-level with the same number of switching components and device topology. Table 6.7 shows the switching states and output voltage levels of two series connected H-bridge units for trinary ratios of DC sources in an AC-MLI. When the DC sources are trinary type, their values are $V_1 = V$, $V_2 = 3V$, $V_3 = 9V$ and $V_4 = 27V$. Trinary ratio in the AC-MLI allows more levels, but the control method of the inverter will be complicated when compared to the binary ratio [6].

$$N_S = n \qquad (6.15)$$

Table 6.7 Switching states for 9-level AC-MLI topology.

V_{out}	Level	\multicolumn{4}{c	}{H-bridge unit-1}	\multicolumn{4}{c}{H-bridge unit-2}
		S_{11}	S_{12}	S_{13}	S_{14}	S_{21}	S_{22}	S_{23}	S_{24}
4V	$(V_1 + V_2)$	1	0	0	1	1	0	1	0
3V	V_2	0	1	0	1	1	0	0	1
2V	$(V_2 - V_1)$	0	1	1	0	1	0	0	1
V	V_1	1	0	0	1	0	1	0	1
0	0	0	0	0	0	0	0	0	0
−V	$-V_1$	0	1	1	0	0	1	0	1
−2V	$-(V_2 - V_1)$	1	0	0	1	0	1	1	0
−3V	$-V_2$	0	1	0	1	0	1	1	0
−4V	$-(V_1 + V_2)$	0	1	1	0	0	1	1	0

$$N_{SW} = 4n \tag{6.16}$$

$$N_{L(binary)} = 2^{(n+1)} - 1 \tag{6.17}$$

$$N_{L(trinary)} = 3^n \tag{6.18}$$

Fig. 6.14 highlights output levels of the AC-MLI in symmetric and asymmetric modes. The asymmetric mode is compared for the binary and trinary type of DC source ratios. While the number of DC sources increases, the output level of the inverter rises due to the ratio of the DC source. The trinary type of asymmetric mode provides the highest level at the output.

6.3.2 Cascaded basic blocks multilevel inverter

Mokhberdoran et al. [19] has proposed a cascaded basic blocks multilevel inverter (CBB-MLI), which is shown in Fig. 6.15. The CBB-MLI topology can be used in symmetric or asymmetric mode, where six switches, eight diodes, and two DC sources are used in a basic unit and the units can be connected in series. A single

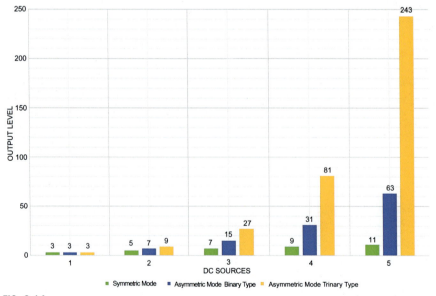

FIG. 6.14

Output levels of AC-MLI in symmetric, asymmetric binary type, and asymmetric trinary type.

6.3 Asymmetric MLI topologies without polarity generation module

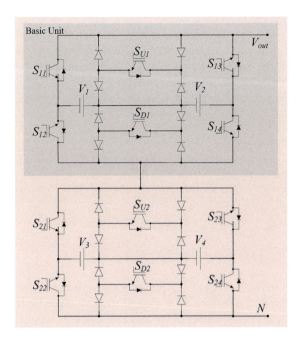

FIG. 6.15

General structure of CBB-MLI.

Table 6.8 Voltage level for one basic unit in asymmetric in CBB-MLI.

V_{out}	Level	S_{11}	S_{12}	S_{13}	S_{14}	S_{U1}	S_{D1}
3V	$V_1 + V_2$	0	1	0	0	0	0
2V	V_2	0	0	1	0	0	1
V	V_1	0	1	1	0	1	0
0	0	0	0	0	1	1	1
		0	0	1	0	0	0
		1	1	0	0	0	0
−V	$-V_1$	1	0	0	0	0	0
−2V	$-V_2$	0	0	0	1	1	0
−3V	$-(V_1 + V_2)$	1	0	0	1	0	0

basic unit can produce a five-level output in symmetric mode or seven-level output in asymmetric mode. Table 6.8 shows the voltage level for one basic unit in asymmetric mode while the DC sources are $V_1 = V$, $V_2 = 2V$ [19].

Different and disordered DC source ratios are proposed in Ref. [19] for asymmetric operation mode. The first proposed ratio is a power of five, where Eq. (6.19) is used for calculating the magnitude of DC sources, with "n" denoting the number of series connected units. The variables and components can be calculated by using Eqs (6.20)–(6.23), where N_S is the number of DC sources, N_{SW} is the number of switching components, N_D is the number of diodes, and N_L is the output level of the inverter in these equations. When the DC sources are used at levels of $V_1 = V$, $V_2 = V$, $V_3 = 5V$, $V_4 = 5V$ in the first proposed ratio, the output waveform is obtained at a 25-level CBB-MLI [19].

$$V_{(2m-1)} = V_{(2m)} = 5^{(m-1)} \cdot V_{DC} \; m = 1,2,\ldots,n \tag{6.19}$$

$$N_S = 2n \tag{6.20}$$

$$N_{SW} = 6n \tag{6.21}$$

$$N_D = 8n \tag{6.22}$$

$$N_L = 5^n \tag{6.23}$$

The second proposed ratio is based on a power of two as in the binary type. Eq. (6.24) is used for calculating the magnitude of the DC sources where "n" is the number of series connected units. The number of DC sources N_S is calculated using Eq. (6.25), while the number of switching components N_{SW} is shown in Eq. (6.26). N_D is the number of diodes and N_L is the output level of the inverter in Eqs (6.27) and (6.28), respectively. When the DC sources are used in binary type in CBB-MLI, their values are determined as $V_1 = V$, $V_2 = 2V$, $V_3 = 4V$, $V_4 = 8V$ in the second proposed ratio for a 31-level output at the CBB-MLI [19].

$$V_{(m)} = 2^{(m-1)} V_{DC} \; m = 1,2,\ldots,2n \tag{6.24}$$

$$N_S = 2n \tag{6.25}$$

$$N_{SW} = 6n \tag{6.26}$$

$$N_D = 8n \tag{6.27}$$

$$N_L = 2^{(2n+1)} - 1 \tag{6.28}$$

The third application of CBB-MLI is based on a ratio of seven, as seen in Eq. (6.29) for calculating the magnitude of DC sources while "n" is the number of series connected units. Variables and components can be calculated by using Eqs (6.30)–(6.33). N_S is the number of DC sources, N_{SW} is the number of switching components, N_D is the number of diodes, and N_L is the output level of the inverter in these equations. When the DC sources are used as $V_1 = V$, $V_2 = 2V$, $V_3 = 7V$, $V_4 = 14V$ levels, the staircase levels of output waveforms are obtained at the 49-level in the CBB-MLI topology [19].

6.3 Asymmetric MLI topologies without polarity generation module

$$V_{(m)} = \begin{cases} 7^{(m-1)/2} V_{DC} \text{ for } m = odd \\ 2 \times 7^{(m-2)/2} V_{DC} \text{ for } m = even \end{cases} m = 1, 2, \ldots, 2n \quad (6.29)$$

$$N_S = 2n \quad (6.30)$$

$$N_{SW} = 6n \quad (6.31)$$

$$N_D = 8n \quad (6.32)$$

$$N_L = 7^n \quad (6.33)$$

6.3.3 Cascaded modified H-bridge multilevel inverter

Babaei et al. suggested a cascade H-bridge based MLI topology in Ref. [20]. This topology is defined as a cascaded modified H-bridge multilevel inverter (CMB-MLI); the basic unit and series connection of units are depicted in Fig. 6.16. Each basic unit comprises six switches and two DC sources. The CMB-MLI topology can be operated in symmetric and asymmetric mode. In this topology, the asymmetric operation mode enables a larger number of levels than symmetric operation. The output voltage levels for a single basic unit are shown in Table 6.8. The authors of

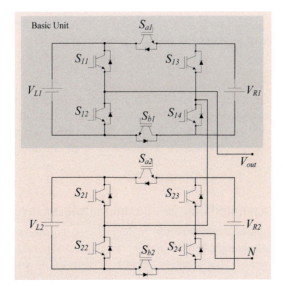

FIG. 6.16

General structure of CMB-MLI.

Table 6.9 Switching states for one basic unit in CBB-MLI.

V_{out}	S_{11}	S_{12}	S_{13}	S_{14}	S_{a1}	S_{b1}
0	1	0	1	0	1	0
0	0	1	0	1	0	1
V_{L1}	1	0	0	1	0	1
V_{R1}	0	1	1	0	0	1
$V_{R1} + V_{L1}$	1	0	1	0	0	1
$-V_{L1}$	0	1	1	0	1	0
$-V_{R1}$	1	0	0	1	1	0
$-(V_{R1} + V_{L1})$	0	1	0	1	1	0

[20] have proposed nine different algorithms for ratios of DC sources. To achieve the highest available number of output levels, the number of series connected units and DC sources (V_R and V_L) are calculated as illustrated in Eqs (6.34) and (6.35). The variables and components can be calculated using Eqs (6.36)–(6.38), where N_S is the number of DC sources, N_{SW} is the number of switching components, and N_L is the output level of the inverter in these equations where the output levels of a single CCB-MLI inverter are seen in Table 6.9.

$$V_R = 7^{(n-1)} V_{DC} \quad (6.34)$$

$$V_L = 2 \times 7^{(n-1)} V_{DC} \quad (6.35)$$

$$N_S = 2n \quad (6.36)$$

$$N_{SW} = 6n \quad (6.37)$$

$$N_L = 7^n \quad (6.38)$$

Fig. 6.17 shows output levels of CMB-MLI in symmetric and asymmetric operation modes. There are two sources and six switches in a single basic unit, so these values are the same for symmetric and asymmetric modes. It can be seen from Fig. 6.17 that the output levels of the inverter in asymmetric mode are much higher than in symmetric mode.

6.3.4 Cross connected sources based multilevel inverter

Gupta and Jail proposed a new topology, called cross connected sources based MLI (CCS-MLI). In this topology, the DC sources are connected to each other in opposite polarities with switches. This topology can also be operated in symmetric and asymmetric modes; the structure of the topology is depicted in Fig. 6.18. The ratio of DC sources is the disordered type in asymmetric mode, where source ratios are not like the order of V, $2V$, $3V$, $4V$. The required parameters and component values can be

6.3 Asymmetric MLI topologies without polarity generation module

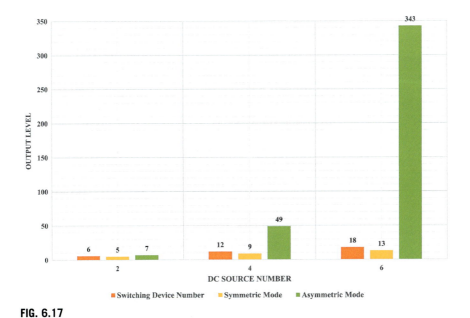

FIG. 6.17

Output levels of CMB-MLI in symmetric and asymmetric modes.

calculated using Eqs (6.39)–(6.43) where N_S is the number of DC sources and the value of the sources is calculated using different methods for even or odd number sources. The $V_{DC(even)}$ and $V_{DC(odd)}$ calculations are depicted in Eqs (6.40) and (6.41), respectively, where N_{SW} is the number of switching components and N_L is the output level of the inverter in these equations [21, 22].

$$N_S = n \tag{6.39}$$

$$V_{DC(even),j} = \begin{cases} (2j-1)V_{DC} \text{ for } 1 \leq j \leq (N_S/2) \\ 2(N_S+1-j)V_{DC} \text{ for } (N_S/2) < j \leq N_S \end{cases} \tag{6.40}$$

$$V_{DC(odd),j} = \begin{cases} (2j-1)V_{DC} \text{ for } 1 \leq j \leq ((N_S+1)/2) \\ 2(N_S+1-j)V_{DC} \text{ for } ((N_S+1)/2) < j \leq N_S \end{cases} \tag{6.41}$$

$$N_{SW} = 2n+2 \tag{6.42}$$

$$N_L = n^2 + n + 1 \tag{6.43}$$

Fig. 6.19 shows the calculated output levels of CCS-MLI in symmetric and asymmetric operation modes for comparison. It can be seen from Fig. 6.19 that the output levels of the inverter in asymmetric mode are higher than in symmetric mode. The switching states of CCS-MLI are illustrated in Table 6.10 for a 13-level A-MLI.

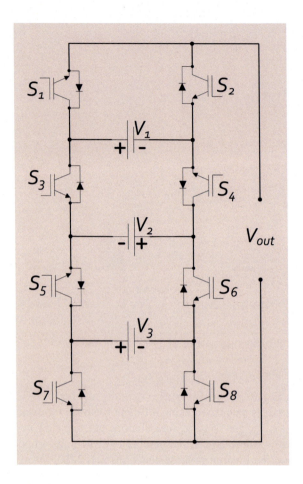

FIG. 6.18

General structure of CCS-MLI.

The DC sources are $V_1 = V_{DC}$, $V_2 = 3V_{DC}$, $V_3 = 2V_{DC}$ as Eqs (6.40) and (6.41), where only sequential sources can be added to each other. For example, V_1 and V_2 can be connected, while V_1 and V_3 cannot be connected in this topology. The configuration does not need an H-bridge structure due to the polarity of the DC sources. Therefore the output level is decreased based on prevention of cascading all available DC sources. The source connection sequences are given in Table 6.11, considering up to 10 sources. The magnitudes of these sources are computed using Eqs (6.40) and (6.41). The topology uses a very complicated source sequence and it requires predefinitions that are assumed to be drawbacks of this topology [21, 22].

6.3 Asymmetric MLI topologies without polarity generation module

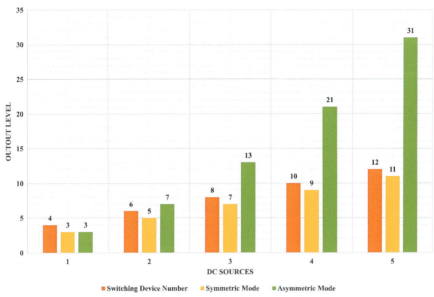

FIG. 6.19

Output levels of CCS-MLI in symmetric and asymmetric mode.

Table 6.10 Switching states for three numbers of DC sources in CCS-MLI.

Level	V_{out}	S_1	S_2	S_3	S_4	S_5	S_6	S_7	S_8
6V	$V_1 + V_2 + V_3$	1	0	0	1	1	0	0	1
5V	$V_2 + V_3$	0	1	0	1	1	0	0	1
4V	$V_1 + V_2$	1	0	0	1	1	0	1	0
3V	V_2	0	1	0	1	1	0	1	0
2V	V_3	1	0	1	0	1	0	0	1
V	V_1	1	0	0	1	0	1	0	1
0	0	0	1	0	1	0	1	0	1
		1	0	1	0	1	0	1	0
$-V$	$-V_1$	0	1	1	0	1	0	1	0
$-2V$	$-V_3$	0	1	0	1	0	1	1	0
$-3V$	$-V_2$	1	0	1	0	0	1	0	1
$-4V$	$-(V_1 + V_2)$	0	1	1	0	0	1	0	1
$-5V$	$-(V_2 + V_3)$	1	0	1	0	0	1	1	0
$-6V$	$-(V_1 + V_2 + V_3)$	0	1	1	0	0	1	1	0

Table 6.11 DC sources sequence in CCS-MLI for 10 sources.

Number of DC sources	V_1	V_2	V_3	V_4	V_5	V_6	V_7	V_8	V_9	V_{10}
1	V_{DC}	X	X	X	X	X	X	X	X	X
2	V_{DC}	$2V_{DC}$	X	X	X	X	X	X	X	X
3	V_{DC}	$3V_{DC}$	$2V_{DC}$	X	X	X	X	X	X	X
4	V_{DC}	$3V_{DC}$	$4V_{DC}$	$2V_{DC}$	X	X	X	X	X	X
5	V_{DC}	$3V_{DC}$	$5V_{DC}$	$4V_{DC}$	$2V_{DC}$	X	X	X	X	X
6	V_{DC}	$3V_{DC}$	$5V_{DC}$	$6V_{DC}$	$4V_{DC}$	$2V_{DC}$	X	X	X	X
7	V_{DC}	$3V_{DC}$	$5V_{DC}$	$7V_{DC}$	$6V_{DC}$	$4V_{DC}$	$2V_{DC}$	X	X	X
8	V_{DC}	$3V_{DC}$	$5V_{DC}$	$7V_{DC}$	$8V_{DC}$	$6V_{DC}$	$4V_{DC}$	$2V_{DC}$	X	X
9	V_{DC}	$3V_{DC}$	$5V_{DC}$	$7V_{DC}$	$9V_{DC}$	$8V_{DC}$	$6V_{DC}$	$4V_{DC}$	$2V_{DC}$	X
10	V_{DC}	$3V_{DC}$	$5V_{DC}$	$7V_{DC}$	$9V_{DC}$	$10V_{DC}$	$8V_{DC}$	$6V_{DC}$	$4V_{DC}$	$2V_{DC}$

6.3.4.1 Modular parallel sources multilevel inverter

Boora et al. [23] proposed an alternative asymmetrical MLI topology. The authors have not provided any particular name for this topology, but it can be defined as a modular parallel source based multilevel inverter (MPS-MLI). The general structure of this topology is shown in Fig. 6.20, where a basic unit comprises a 7-level asymmetrical MLI. The switching sequence is given in Table 6.12 and the DC sources are integrated into the topology in parallel, as seen in Fig. 6.21. The ratio of DC sources is not ordered in a standard type in the asymmetric operation mode, and the ratios of DC sources do not increment, respectively. The variables and components can be calculated using Eqs (6.44)–(6.47), where N_S denotes the number of DC sources and the value of sources is calculated in different ways for even or odd numbered sources.

$$N_S = n \quad (6.44)$$

$$V_j = \begin{Bmatrix} 5^{\left(\frac{j-1}{2}\right)} V_{DC} \text{ for } j = odd \\ 2 \times 5^{(j/2)-1} V_{DC} \text{ for } j = even \end{Bmatrix} \quad (6.45)$$

$$N_{SW} = 2n + 2 \quad (6.46)$$

$$N_L = 2^{n+1} - 1 \quad (6.47)$$

The values of sources in the MPS-MLI topology can be calculated as shown in Eq. (6.45), where N_{SW} is the number of switching components and N_L is the output

6.3 Asymmetric MLI topologies without polarity generation module

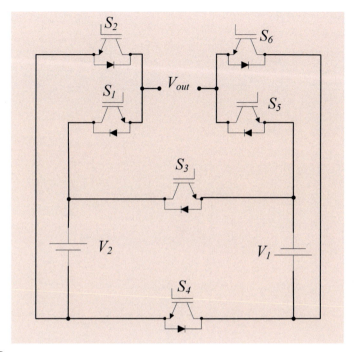

FIG. 6.20

General structure of MPS-MLI.

Table 6.12 Switching sequence of basic MPS-MLI unit.

Level	V_{out}	S_1	S_2	S_3	S_4	S_5	S_6
3V	$V_1 + V_2$	1	0	0	1	1	0
2V	V_2	1	0	0	1	0	1
V	V_1	0	1	0	1	1	0
0	0	0	1	0	1	0	1
−V	$-V_1$	1	0	1	0	0	1
−2V	$-V_2$	0	1	1	0	1	0
−3V	$-(V_1 + V_2)$	0	1	1	0	0	1

level of the inverter in these equations. The switching sequence of a 15-level MPS-MLI is given in Table 6.13 where the output levels are achieved by the sum or subtract of DC sources of $V_1 = V_{DC}$, $V_2 = 2V_{DC}$, $V_3 = 5V_{DC}$ in 15-level MPS-MLI. The odd and even levels are incremented as a power of five while it is increased twice of odd levels in even arrangement [23].

210 CHAPTER 6 Asymmetrical multilevel inverter topologies

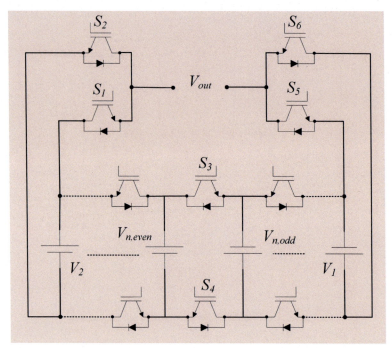

FIG. 6.21

Extension structure of MPS-MLI.

Table 6.13 Switching sequence 15-level MPS-MLI.

Level	V_{out}	S_1	S_2	S_3	S_4	S_5	S_6	S_7	S_8
7V	$V_2 + V_3$	1	0	0	1	1	0	1	0
6V	$V_2 + V_3 - V_1$	1	0	0	1	0	1	1	0
5V	V_3	0	1	0	1	1	0	1	0
4V	$V_3 - V_1$	0	1	0	1	0	1	1	0
3V	$V_1 + V_2$	1	0	0	1	1	0	0	1
2V	V_2	1	0	0	1	0	1	0	1
V	V_1	0	1	0	1	1	0	0	1
0	0	0	1	0	1	0	1	0	1
		1	0	1	0	1	0	1	0
$-V$	$-V_1$	1	0	1	0	0	1	1	0
$-2V$	$-V_2$	0	1	1	0	1	0	1	0
$-3V$	$-(V_1 + V_2)$	0	1	1	0	0	1	1	0
$-4V$	$-(V_3 - V_1)$	1	0	1	0	1	0	0	1
$-5V$	$-V_3$	1	0	1	0	0	1	0	1
$-6V$	$-(V_3 + V_2 - V_1)$	0	1	1	0	1	0	0	1
$-7V$	$-(V_3 + V_2)$	0	1	1	0	0	1	0	1

6.4 Remarks and conclusion

This chapter presents the novel A-MLI topologies, which have recently been proposed in the literature. The common effect of A-MLI topologies is that they can produce more levels with the same number of components as compared with symmetric MLI topologies. These topologies can be classified as symmetric or asymmetric based on the ratio of DC sources supplying the inverter. The generation of an increased number of output waveforms by using a minimum number of components is a crucial topic in inverter topology enhancements. Most of the A-MLI topologies use only switching components in their structure. Therefore, semiconductor devices are the important components in such topologies. Each semiconductor needs a driver circuit, and therefore reducing the number of semiconductor devices used decreases the cost of the inverter in terms of switching and driver circuit requirements. Fig. 6.22 is given in order to illustrate the number of switching components in different DC source numbers and different inverter topologies. The SCH-MLI, CBB-MLI, and CMB-MLI topologies need an increased number of switching devices comparing to other types of MLI topologies. The SA-MLI topology is the best topology in terms of the reduced number of switching devices. However, this is not the only parameter for choosing an inverter topology. Additionally, the AC-MLI and CMB-MLI topologies have different ratios of DC sources in the same structure and same number of switching devices.

The output level is the second important parameter for determining the A-MLI topology to be used. Table 6.14 shows the output level at different DC source numbers for A-MLI topologies. The topologies are sorted from the minimum level to maximum level, where the CCS-MLI topology has the minimum output level and the SCH-MLI topology provides the maximum output level against an increased number of DC sources. The SCH-MLI generates output waveforms in 125 levels by using 3 DC sources, while CCS MLI achieves 111 levels of outputs when it is supplied by 10 separate DC sources. However, Fig. 6.22 shows that SCH-MLI needs a high number of switching devices for generating these output levels. Therefore, the SCH-MLI topology is much more appropriate for use in the decreased number of DC sources. When it is compared according to the minimum number of switching devices that ensures the lowest cost for the structure, the SA-MLI topology would be the most appropriate structure to be used.

Some of the topologies use additional components as diode or capacitors. The SA-MLI and SCH-MLI topologies require only one diode for each DC source, while CBB-MLI uses eight diodes for every basic unit. Only the SCH-MLI topology uses a capacitor in the structure among other topologies compared. Therefore the highest number of components is used in CBB-MLI topology. For example, it will be possible to generate 125-level output by using 3 DC sources and 18 switching devices in SCH-MLI topology, which would be capable of generating the highest number of output levels. However, if the use of a minimum number of switching devices is required, an SA-MLI that is capable of generating 15 levels with 7 switching devices would be chosen.

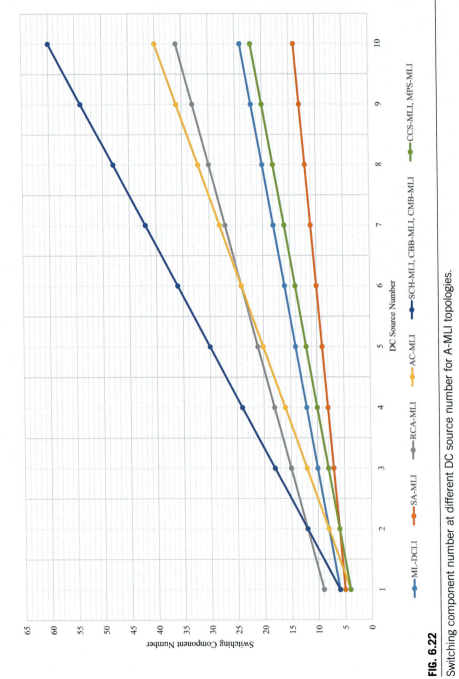

FIG. 6.22

Switching component number at different DC source number for A-MLI topologies.

Table 6.14 Output level at different DC source number for A-MLI topologies.

A-MLI topology	DC source number									
	1	2	3	4	5	6	7	8	9	10
CCS-MLI	3	7	13	21	31	43	57	73	91	111
MPS-MLI	a	5	9	17	33	65	129	257	513	1025
SA-MLI, AC-MLI (Binary)	3	7	15	31	63	127	255	511	1023	2047
ML-DCLI	5	9	17	33	65	129	257	513	1025	2049
CBB-MLI, CMB-MLI	a	7	a	49	a	343	a	2401	a	16,807
RCA-MLI	3	9	25	67	177	465	1219	3193	8361	21,891
AC-MLI (Trinary)	3	9	27	81	243	729	2187	6561	19,683	59,049
SCH-MLI	5	25	125	625	3125	15,625	78,125	390,625	1,953,125	9,765,625

[a] =Invalid structure. Topology needs minimum two sources or increment of sources minimum two.

References

[1] I.H. Shanono, N.R.H. Abdullah, A. Muhammad, A survey of multilevel voltage source inverter topologies, controls and applications, Int. J. Power Elec. Drive Systems 9 (2018) 1186–1201, https://doi.org/10.11591/ijpeds.v9n3.pp1186-1201.

[2] J. Venkataramanaiah, Y. Suresh, A.K. Panda, A review on symmetric, asymmetric, hybrid and single DC sources based multilevel inverter topologies, Renew. Sustain. Energy Rev. 76 (2017) 788–812, https://doi.org/10.1016/j.rser.2017.03.066.

[3] I. Colak, E. Kabalci, R. Bayindir, Review of multilevel voltage source inverter topologies and control schemes, Energ. Conver. Manage. 52 (2011) 1114–1128, https://doi.org/10.1016/j.enconman.2010.09.006.

[4] K.K. Gupta, A. Ranjan, P. Bhatnagar, L.K. Sahu, S. Jain, Multilevel inverter topologies with reduced device count: a review, IEEE Trans. Power Electron. 31 (2016) 135–151, https://doi.org/10.1109/TPEL.2015.2405012.

[5] P. Kala, S. Arora, A comprehensive study of classical and hybrid multilevel inverter topologies for renewable energy applications, Renew. Sustain. Energy Rev. 76 (2017) 905–931, https://doi.org/10.1016/j.rser.2017.02.008.

[6] İ. Çolak, E. Kabalcı, G. Keven, Asimetrik Çok Seviyeli Eviricilerin İncelenmesi/a review on asymmetric multi-level inverters, EMO Bilimsel Dergi. 2 (2012) 29–36.

[7] G.-J. Su, Multilevel DC-Link Inverter, in: *Conference Record of the* 2004 *IEEE Industry Applications Conference*, 2004. *39th IAS Annual Meeting*, 41, 2005, pp. 848–854, https://doi.org/10.1109/IAS.2004.1348506.

[8] N. Sivakumar, A. Sumathi, R. Revathy, THD analysis of a 13 level asymmetric hybrid cascaded multilevel inverter for industrial applications, World Eng. Appl. Sci. J 6 (2015) 109–118, https://doi.org/10.5829/idosi.weasj.2015.6.2.22150.

[9] M.A.N. Doss, R. Naveenkumar, R. Ravichandran, J. Rengaraj, M. Manikandan, PV fed asymmetrical switched diode multi level inverter with minimum number of power electronic components, Energy Procedia 117 (2017) 592–599, https://doi.org/10.1016/j.egypro.2017.05.155.

[10] J. Madhusudhana, P.S. Puttaswamy, Genetic algorithm based 15 level modified multilevel inverter for stand alone photovoltaic applications, Int. J. Mod. Trends Eng. Res. 3 (2016) 355–367.

[11] C. Bharatiraja, H. Reddy, N. Sri Ramsai, S.S. Saisuma, FPGA based design and validation of asymmetrical reduced switch multilevel inverter, Int. J. Power Electron. Drive Syst. 7 (2016) 340–348, https://doi.org/10.11591/ijpeds.v7.i2.pp340-348.

[12] J.G. Shankar, Design and implementation of 15-level asymmetric cascaded H bridge multilevel inverter, J. Electr. Eng. 17 (2018) 396–404.

[13] D.D. Naidu, D.V. Kumar, A. New Simplified, Asymmetrical multilevel inverter topology with fundamental switching control, IJERGS 4 (2016) 473–479.

[14] E. Babaei, S.S. Gowgani, Hybrid multilevel inverter using switched capacitor units, IEEE Transactions on Industrial Electronics 61 (2014) 4614–4621, https://doi.org/10.1109/TIE.2013.2290769.

[15] M.S. Bin Arif, S.M. Ayob, A. Iqbal, S. Williamson, Z. Salam, Nine-level asymmetrical single phase multilevel inverter topology with low switching frequency and reduce device counts, in: Proceedings of the IEEE International Conference on Industrial Technology, 2017, pp. 1516–1521, https://doi.org/10.1109/ICIT.2017.7915591.

[16] M.S. Bin Arif, S.M. Ayob, Z. Salam, Asymmetrical multilevel inverter topology with reduced power semiconductor devices, in: *IEAC on 2016–2016 IEEE Industrial Electronics and Applications Conference*, 2017, pp. 20–25, https://doi.org/10.1109/IEACON.2016.8067349.

[17] M.S. Arif, S.M. Ayob, Z. Salam, Asymmetrical nine-level inverter topology with reduce power semicondutor devices, Telkomnika 16 (2018) 38–45, https://doi.org/10.12928/TELKOMNIKA.v16i1.8520.

[18] S. Kouro, M. Malinowski, K. Gopakumar, J. Pou, L.G. Franquelo, B. Wu, J. Rodriguez, M.A. Perez, J.I. Leon, Recent advances and industrial applications of multilevel converters, IEEE Transactions on Industrial Electronics. 57 (2010) 2553–2580, https://doi.org/10.1109/TIE.2010.2049719.

[19] A. Mokhberdoran, A. Ajami, Symmetric and asymmetric design and implementation of new cascaded multilevel inverter topology, IEEE Trans. Power Electron. 29 (2014) 6712–6724, https://doi.org/10.1109/TPEL.2014.2302873.

[20] E. Babaei, S. Laali, S. Alilu, Cascaded multilevel inverter with series connection of novel H-bridge basic units, IEEE Transactions on Industrial Electronics 61 (2014) 6664–6671, https://doi.org/10.1109/TIE.2014.2316264.

[21] K.K. Gupta, S. Jain, A novel multilevel inverter based on switched dc sources, IEEE Transactions on Industrial Electronics 61 (2014) 3269–3278, https://doi.org/10.1109/TIE.2013.2282606.

[22] K.K. Gupta, S. Jain, Comprehensive review of a recently proposed multilevel inverter, IET Power Electron. 7 (2014) 467–479, https://doi.org/10.1049/iet-pel.2012.0438.

[23] K. Boora, J. Kumar, A new general topology for asymmetrical multilevel inverter with reduced number of switching components, içinde: 2017, in: Recent Developments in Control, Automation & Power Engineering (RDCAPE), IEEE, 2017, pp. 66–71.

CHAPTER 7

Resonant and Z-source multilevel inverters

Oleksandr Husev[a,b] and Carlos Roncero-Clemente[c,d]

[a]Tallinn University of Technology, Department of Mechatronics and Electrical Engineering, Tallinn, Estonia [b]Department of Electrical Power Engineering and Mechatronics, TalTech University, Tallinn, Estonia [c]Power Electrical and Electronic Systems Research Group, University of Extremadura, Badajoz, Spain [d]Electrical, Electronic and Control Engineering Department, Extremadura University, Badajoz, Spain

List of abbreviations

ANPC	active NPC
ARCP	auxiliary resonant commutated poles
CCM	continuous conduction mode
CHB	cascaded H-bridge
CIC	continuous input current
CSI	current-source inverter
CSN	controlled switch network
DCLC	DC-Link cascaded
DCM	discontinuous conduction mode
DIC	discontinuous input current
EMI	electromagnetic interferences
C	flying capacitor
IS	impedance source
LPF	low-pass filter
MLI	multilevel inverters
PWM	pulse width modulation
qZS	quasi-Z-source
PRC	resonant circuits
RTN	resonant tank network
SRC	series-resonant
ST	shoot-though
VSI	voltage-source inverter
ZCS	zero current switching
ZS	Z-source
ZSI	Z-source inverter
ZV	zero-voltage
ZVS	zero voltage switching

7.1 General operating principle of resonant circuits

There are some constraints describing the switch-mode operation of a wide variety of pulse width modulated (PWM) DC-DC and DC-AC power converters. First, during the real switch-on and switch-off times, a high current and/or voltage may appear in and across the power semiconductors, contributing to a higher power loss and stress that linearly increases as the switching frequency increases. This increase is desired in order to reduce the passive element sizes, which leads to a higher power density inss the power converter. Second, another relevant disadvantage is the large di/dt and dv/dt produced during the switching performance, increasing the electromagnetic interference (EMI).

Thus with the main aim of increasing the switching frequency and avoiding the aforementioned drawbacks, the resonant converters appeared [1]. The core idea of this family of converters relies on realizing the switching when the voltage across a power switch and/or the current through it is zero during the switching time. In this way, a simultaneous voltage and current transition is avoided, thereby eliminating the switching power losses. This phenomenon is achieved by a proper switching strategy [2], which provides the so-called soft switching (instead of hard switching in the PWM converters). Roughly, the power converter includes a resonant L-C circuit with a certain tuned resonant frequency [3]. By controlling the switching frequency [4–7], the output power is regulated. The soft switching can be produced as zero current switching (ZCS) and/or zero voltage switching (ZVS), depending on whether the current or the voltage is kept at zero during commutations.

It is possible to classify the soft-switching power converters based on their fundamental operating principle and their circuitries. The features of these families are summarized in Table 7.1 and some examples of the topologies are shown in Fig. 7.1.

The family of resonant power converters is becoming more and more popular because of their superior performance, with some emerging applications, such as

Table 7.1 Classification of the resonant converters.

Family	Operating principle	References
Quasiresonant converters and multiresonant	Include two reactive elements (L and C) around the power switches in a hard switching power converter to achieve ZCS or ZVS. Simultaneous ZCS and ZVS is achieved by using three reactive elements	[8–11]
Resonant-transition converters	Include two reactive elements and two power switches as auxiliary circuit to produce the ZCS or ZVS	[12–14]
Resonant power converters	Include two or more reactive elements in cascade with the converter	[15–18]

7.1 General operating principle of resonant circuits

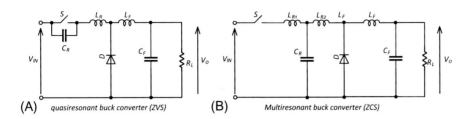

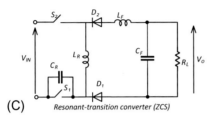

FIG. 7.1

Basic soft-switching converter: quasiresonant buck converter (ZVS) (A); multiresonant buck converter (ZCS) (B); and resonant-transition converter (ZCS) (C).

power supplies and grid-connected inverters in renewable energies, including photovoltaics and wind, fuel cells, and induction cookers. The topology of this system is spread into different stages. The first stage is composed of a controlled switch network (CSN), which generates a high-frequency AC waveform (half- and full-bridge inverters are commonly used for this stage) (Fig. 7.2B). The second stage develops the resonance conditions using a resonant tank network (RTN) in order to provide soft-switching capabilities to the CSN by tuning the voltage and current oscillations of the RTN [19]. Finally, a low-pass filter (LPF) is usually connected to supply the load. Depending on the different arrangements of the RTN and the output filters, different resonant power converters are proposed with different features. The RTN can be composed of two, three, four, or five elements combined in different ways to finally obtain several hundreds of topologies. The higher the number of elements in the RTN, the higher the efficiency obtained. The study of resonant power converters is usually more complex than that for PWM converters in the time domain; thus the fundamental harmonic approximation is the technique adopted to analyze them in the frequency domain. This approach ignores the harmonic (filtered out by the output filter) and the voltage and currents in the RTN are assumed sinusoidal. A general topological scheme is represented in Fig. 7.2A, with an output stage composed of a diode rectifier bridge and the output filter (Fig. 7.2C). In many RPCs, a high-frequency transformer is included to assure a higher voltage gain or galvanic isolation.

220 CHAPTER 7 Resonant and Z-source multilevel inverters

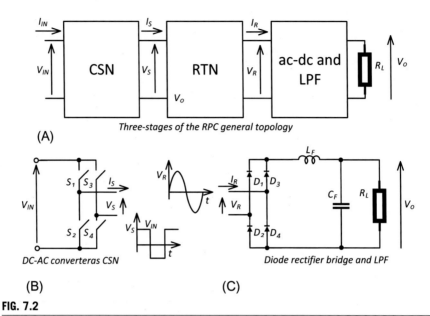

FIG. 7.2
Details of the resonant power converter: (A) different stages of a general RPC topology); (B) full bridge as CSN; and (C) diode rectifier bridge with LPF.

The CSN that composed the first stage will draw a DC current from the input source, generating a square-wave voltage V_S at a certain frequency. The current I_S is sinusoidal due to the filtering action of the RTN and V_R has the same frequency as the input current. The main expressions that characterize the preceding circuit with the capacitor filter are as follows:

$$i_R(t) = I_P \sin(w_S t - \phi_S), \tag{7.1}$$

$$V_{R1} = \frac{4V_0}{\pi} \sin(w_S t - \phi_S), \tag{7.2}$$

$$I_0 = \frac{2}{T_S} \int_0^{T_S/2} i_R(t) dt = \frac{2}{\pi} I_R. \tag{7.3}$$

At the same time, the main parameters (resonant angular frequency (ω_0), characteristic impedance of the resonant circuit (Z_0), and the load quality factor at the resonant frequency for both the series-resonant (SRC) and parallel-(PRC) resonant circuits (Q_L)) are expressed as:

$$w_0 = 2\pi f_0 = \frac{1}{\sqrt{L_R C_R}}, \tag{7.4}$$

7.1 General operating principle of resonant circuits

$$Z_0 = \sqrt{\frac{L_R}{C_R}} = w_0 L_R = \frac{1}{w_0 C_R}, \quad (7.5)$$

$$\begin{cases} Q_L(SRC) = \dfrac{Z_0}{R_0} \\ Q_L(PRC) = \dfrac{R_0}{Z_0} \end{cases} \quad (7.6)$$

One of the basic topologies is the full-bridge series-loaded DC/DC converter, including snubber capacitors parallel to the active power semiconductors. This circuit is displayed in Fig. 7.3. In this case the CSN is composed of a full-bridge arrangement together with the capacitor snubbers. The RTN consists of a series association of L_R and C_R, and the third stage is formed by a diode bridge, and a simple first-order low-pass filter (C_F) is selected to filter out the high-frequency harmonics at the load. The snubber capacitors are charged by the resonant tank current and allow for zero voltage turn-off commutations of the switches.

In this RPC the resonant frequency is around 5 kHz and operates in continuous conduction mode (CCM). The main simulated waveforms are displayed in Fig. 7.4.

Some applications require isolation between the input source and the output. Another well-known topology that belongs to this family is the isolated DC-DC LLC resonant converter (Fig. 7.5) operating with frequency control. The load voltage is controlled by regulating the switching frequency of the H-bridge. The ZVS technique allows reducing switching losses, permitting the performance at higher switching frequencies.

The input source is connected with a front-end full bridge and its AC terminal connects to the RTN via a series-connected resonant L and C, that together with the magnetizing inductance of the transformer compose the RTN (LLC type). The secondary side of the transformer is connected through a full-wave rectifier to supply the load by a third-order filter. This LLC converter works by the ZVS technique, where each power switch (e.g., S_1) is switched on when the current is still circulating across its antiparallel diode. Therefore, the turn-on loss is substantially reduced as the voltage applied to the power switch is the forward voltage drop of the diode, which is

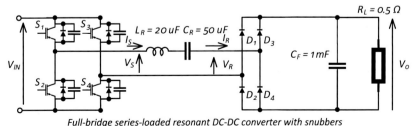

Full-bridge series-loaded resonant DC-DC converter with snubbers

FIG. 7.3

Full-bridge series-loaded resonant DC-DC converter with snubbers.

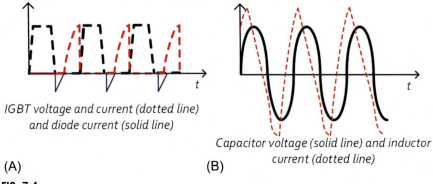

(A) IGBT voltage and current (dotted line) and diode current (solid line)

(B) Capacitor voltage (solid line) and inductor current (dotted line)

FIG. 7.4

Main waveforms of the full-bridge series-loaded resonant DC-DC converter with snubber capacitor: (A) IGBT voltage, IGBT current, and diode current; (B) inductor current and capacitor voltage in the RTN.

very small if compared with the input voltage. The power switch is switched off when it is conducting current, so hard switching is produced as well as turn-off losses.

During the turn-off and turn-on transitions, the body capacitor of the power semiconductor is charged and discharged. A great advantage of ZVS if using FETs, besides the elimination of the turn-on loss, is that the energy stored in the body capacitance of the power switch can be recirculated to the main circuit if the dead-time is sufficient. In this example, a 200 V input voltage is connected to supply the power converter and Fig. 7.6A shows the voltage across a power switch as well as the current flowing through it and the gate signal. The main waveforms of the RTN are displayed in Fig. 7.6B. In this figure we can observe the capacitor voltage, the inductor current, and the primary winding voltage in each element that makes up the resonant tank.

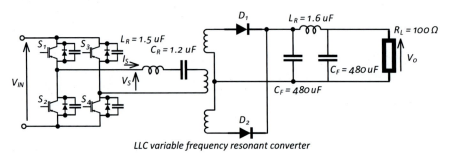

LLC variable frequency resonant converter

FIG. 7.5

Full-bridge series-loaded resonant DC-DC converter with snubbers.

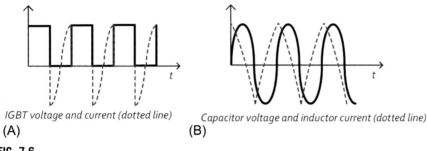

FIG. 7.6

Waveforms at the CSN (IGBT voltage, IGBT current) (A) and waveforms at the RTN: capacitor voltage and inductor current (B).

7.2 General operating principle of impedance-source networks

The Z-source inverter (ZSI) was introduced for the first time in 2003 [20]. This proposed converter overcame the conceptual and theoretical barriers and limitations of the traditional voltage-source inverter (VSI) and current-source inverter (CSI) and provided a novel power conversion concept. The ZSI utilizes shoot-through (ST) cross-conduction states to boost the input DC voltage by switching on both the top and bottom switches of at least one inverter leg. The ZSI can buck-boost voltage, minimize component count, increase efficiency, and reduce cost. Also, this topology does not have any forbidden switching states, which significantly improves converter reliability.

This solution has been intended for various fields of application: DC-DC, AC-DC, AC-AC, DC-AC. In particular it is suitable for grid integration or electric drive control [20–23]. Further, the quasi-Z-source (qZS) network has been proposed in [21].

At the present moment over 20 different types of impedance-source (IS) networks have been presented [24–26]. All of them can be subdivided into several groups. They may have separated inductors, magnetically coupled inductors, or transformers. Implementation of magnetically coupled inductors or transformers in an IS network can result in a higher voltage boost factor due to the turns ratio. Also, all IS networks can be divided into those with discontinuous input current (DIC) and continuous input current (CIC). At the same time, it has been shown that networks with DIC have no advantage over topologies that use CIC [27].

Fig. 7.7 shows basic IS networks. The first Z-source (ZS) network is presented in Fig. 7.7A.

The next IS network under consideration is the qZS network (Fig. 7.7B). As was mentioned earlier, it has CIC and the same size and volume of passive components as ZSI. The next four selected topologies belong to the magnetically coupled IS networks [28, 29]. Since any magnetically coupled inductor can be represented as a combination of leakage inductance, magnetizing inductance, and ideal transformer,

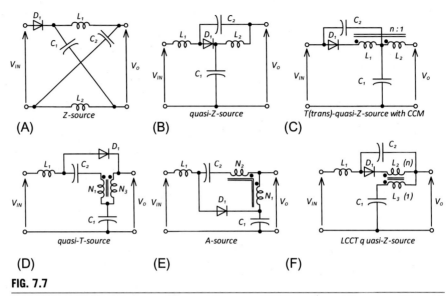

FIG. 7.7

Basic IS networks: (A) Z-source; (B) quasi-Z-source; (C) T(trans)-quasi-Z-source; (D) quasi-T-source; (E) A-source; and (F) LCCT quasi-Z-source.

any other IS network with a magnetically coupled inductor can be considered to be a derivation of the solutions presented. This was very well demonstrated in [30].

The T(trans)-quasi-Z-source network (Fig. 7.7C) with CIC is the first network with coupled inductors under consideration [31–37]. As was shown in [38], despite an additional capacitor the overall size of the capacitors is lower. The quasi-T-source network (Fig. 7.7D) has CIC and a slightly different configuration.

LCCT networks (Fig. 7.7F) have quite similar features [39]. At the same time, there are many derivative circuits and types of converters proposed [40–43]. Finally, the A-source network (Fig. 7.7E) is one of the latest proposed solutions and will be compared as well [44, 45].

The main advantage of all magnetically coupled derived IS networks is the high boost due to the high turns ratio of the transformer. As a result, the ST duty cycle is shorter. At the same time, there are no papers that have strictly demonstrated the impact of the shorter ST duty cycle on the size, volume, and probably cost of the converter.

Generalization of these networks is given in several works [24–26]. As was mentioned earlier, IS networks can be subdivided into the following groups: those including inductive components, magnetically coupled inductive components, and a transformer. Implementation of magnetically coupled transformers in the IS networks can result in a higher voltage boost factor due to the turns ratio. Table 7.2 shows classification of the IS networks [26].

7.2 General operating principle of impedance-source networks

Table 7.2 Classification of the ISNs.

	Separate inductors	Coupled inductors	Transformers
DIC	Z-source	T(trans)-Z-source Y-source TZ-source	LCCT-Z-source Γ-Z-source
CIC	Quasi-Z-source EZ-source	EZ-source T(trans)-quasi-source T(trans)-quasi-source with CCM	LCCT quasi-Z-source

The two possible applications of any IS network are shown in Fig. 7.8. In the case of DC-DC applications, the DC-link voltage is close to the peak voltage across the IS network (Fig. 7.8A). At the same time, the DC-AC converter based on the IS network does not have a direct DC link. When the load terminals are shorted through both the upper and lower semiconductor devices of any one phase leg, the energy accumulates in the inductors of the IS network.

This ST zero state provides the unique buck-boost feature to the inverter. At the same time, the average voltage applied to the AC load is lower than the peak voltage across the IS network. The ST duty cycle inserted in the switching states reduces effective DC-link voltage. In other words, the peak voltage generated from the IS network across the inverter should be higher than in a conventional VSI in order to compensate for the zero ST states (Fig. 7.8B). This has to be taken into account during design of the converter. In particular, voltage stress across the semiconductors is increasing.

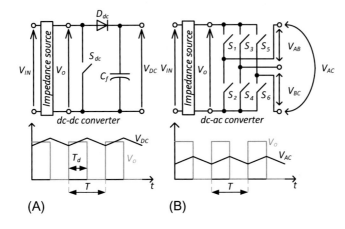

FIG. 7.8

Output voltage of (A) impedance source DC-DC converter and (B) impedance source DC-AC converter.

Fig. 7.9 shows equivalent circuits of the ZS and qZS networks. It shows the two main states of the DC-DC or DC-AC converter that are based on the ZS or qZS network. The ST state (Fig. 7.9A and C) corresponds to the accumulating energy time period, which the duty cycle usually denotes as DS.

During the ST time interval, the current in the inductor is increasing and energy is accumulating in the inductors. The non-ST equivalent circuit state is depicted in Fig. 7.9B and D. It corresponds to the time interval when energy is provided to the load and charges the capacitors while current is decreasing. The average current value remains the same in the steady-state condition. The inductance value is usually selected to limit this current ripple. This value of inductance along with peak current value defines the size, volume, and finally the cost of the inductors.

Steady-state analysis consists of the conditions wherein average voltage across inductance and average current across capacitors are equal to zero [46].

For the ST time interval for a ZS network, the following differential equations are valid:

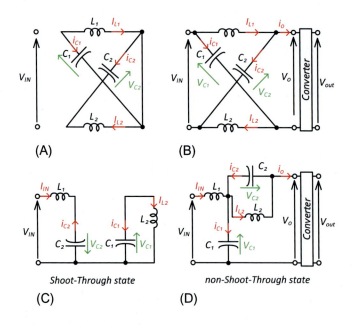

FIG. 7.9

Equivalent circuits of ZS and qZS networks: (A) ST state of ZS network and (B) non-ST state of ZS network; (C) ST state of qZS network and (D) non-ST state of qZS network.

7.2 General operating principle of impedance-source networks

$$\begin{cases} L_1 \dfrac{dI_{L1}}{dt} = v_{C1} \\ L_2 \dfrac{dI_{L2}}{dt} = v_{C2} \\ C_1 \dfrac{dV_{C1}}{dt} = -I_{L1} \\ C_2 \dfrac{dV_{C2}}{dt} = -I_{L2} \end{cases}, \quad (7.7)$$

For the ST time interval for a qZS network, similar differential equations are valid:

$$\begin{cases} L_1 \dfrac{dI_{IN}}{dt} = v_{IN} + v_{C2} \\ L_2 \dfrac{dI_{L2}}{dt} = v_{C1} \\ C_1 \dfrac{dV_{C1}}{dt} = -I_{IN} \\ C_2 \dfrac{dV_{C2}}{dt} = -I_{L2} \end{cases}, \quad (7.8)$$

For a non-ST time interval for a ZS network, the following differential equations are valid:

$$\begin{cases} L_1 \dfrac{dI_{L1}}{dt} + L_2 \dfrac{dI_{L2}}{dt} = v_{IN} - v_0 \\ L_1 \dfrac{dI_{L1}}{dt} - L_2 \dfrac{dI_{L2}}{dt} = v_{C1} - v_{C2} \\ C_1 \dfrac{dV_{C1}}{dt} = I_{L1} - I_0 \\ C_2 \dfrac{dV_{C2}}{dt} = I_{L2} - I_0 \end{cases}, \quad (7.9)$$

For a non-ST time interval for a qZS network, the following differential equations are valid:

$$\begin{cases} L_1 \dfrac{dI_{IN}}{dt} = v_{IN} - v_{C1} \\ L_1 \dfrac{dI_{IN}}{dt} + L_2 \dfrac{dI_{L2}}{dt} = v_{IN} - v_0 \\ C_1 \dfrac{dV_{C1}}{dt} - C_2 \dfrac{dV_{C2}}{dt} = I_{IN} - I_{L2} \\ C_1 \dfrac{dV_{C1}}{dt} = I_{IN} - I_0 \end{cases}, \quad (7.10)$$

The steady-state conditions for both circuits are:

$$\langle V_{L1} \rangle_T = \left\langle L_1 \dfrac{dI_{L1}}{dt} \right\rangle_T = 0, \quad \langle V_{L2} \rangle_T = \left\langle L_2 \dfrac{dI_{L2}}{dt} \right\rangle_T = 0,$$

$$\langle I_{C1} \rangle_T = \left\langle C_1 \dfrac{dV_{C1}}{dt} \right\rangle_T = 0, \quad \langle I_{C2} \rangle_T = \left\langle C_2 \dfrac{dV_{C2}}{dt} \right\rangle_T = 0, \quad (7.11)$$

where T is a switching period.

Assuming a lossless system, the power balance for both circuits can be expressed as:

$$\langle P_{IN} \rangle_T = \langle P_{OUT} \rangle_T, \quad (7.12)$$

where P_{IN} is an input power and P_{OUT} is an output power.

For a ZS network it is expressed as:

$$\langle V_{IN} \cdot I_{IN} \rangle_T = \langle V_0 \cdot {}^0 I_0 \rangle_T,$$
$$V_{IN} \langle I_{IN} \rangle_{(1-D)T} (1-D) = \langle V_0 \rangle_{(1-D)T} \cdot \langle I_0 \rangle_{(1-D)T} (1-D), \quad (7.13)$$

where D is ST duty cycle, which is defined as:

$$D = \langle d \rangle_T = \frac{T_d}{T} \quad (7.14)$$

where T is a switching period and T_d is a duration of the ST state.

For a qZS network, the power balance condition has a difference, which is explained by permanent input voltage source connection to the network:

$$\langle V_{IN} \cdot I_{IN} \rangle_T = \langle V_0 \cdot {}^0 I_0 \rangle_T,$$
$$V_{IN} \langle I_{IN} \rangle_T = \langle V_0 \rangle_{(1-D)T} \cdot \langle I_0 \rangle_{(1-D)T} (1-D). \quad (7.15)$$

Taking into account Eqs (7.7), (7.9), and (7.11)–(7.13), the following expressions can be obtained for a Z-source network [20]:

$$B = \frac{\langle V_{DC} \rangle_T}{\langle V_{IN} \rangle_T} = \frac{V_0}{\langle V_{IN} \rangle_T} = \frac{1}{1 - 2 \cdot D}, \quad (7.16)$$

where B is a boost factor in the case of a DC application, which shows the ratio between average output voltage V_{DC} and average input voltage V_{IN}. V_0 is a peak voltage across a Z-source network. Neglecting the high-frequency ripple, it is assumed that peak voltage across a Z-source network is equal to the average DC-link voltage. This assumption is valid if filtering DC-link capacitor C_f (Fig. 7.8) is large enough to minimize the voltage ripple.

The average voltage across ZS capacitors can be expressed as:

$$\langle V_{C1}{}^{C1} \rangle_T = \langle V_{C2}{}^{C2} \rangle_T = \left(\frac{1-D}{1-2 \cdot D} \right) \cdot \langle V_{IN} \rangle_T. \quad (7.17)$$

In the case of a DC-AC application, the voltage across a ZS network defines the output voltage of the inverter. The inverter control system is featured by the use of two carrier signals. Pulse width modulation (PWM) schemes for the ZS inverters are developed from the classical PWM modulation approach. However, correct integration of the ST state sequence with the classical PWM is essential for proper ZS operation, as some of the vectors can cause a short circuit across the full DC link, which then results in zero voltage output. It is important to maintain a normalized volt-second balance while sequencing the ST states, to accurately reproduce the desired three-phase sinusoidal voltages. The control methods are classified here [47, 48].

Detailed analysis of modulation techniques or ST control methods for the three-phase two-level Z-source inverters can be found here [49–55]. These controls are

7.2 General operating principle of impedance-source networks

mainly classified as: simple boost control (SBC) [49], maximum boost control (MBC) [50], maximum constant boost control (MCBC) [51], and modified space vector modulation control (MSVMBC) [52]. In addition to other techniques, they have been used to control converters in such applications as PV solar energy and fuel cells [53, 54].

Also, careful integration of ST with the conventional switching sequences is required to achieve maximum voltage boost, minimum harmonic distortion, lower semiconductor stress, and a minimum number of device commutations per switching cycle. The boost control, or simple boost, is mostly used. In this case, the ST states are equally distributed during all fundamental periods. Thus maximum average output voltage will be equal to the voltage across the capacitors:

$$\langle V_0 \rangle_T = \langle V_{C1}{}^{C1} \rangle_T = \langle V_{C2}{}^{C2} \rangle_T = \left(\frac{1-D}{1-2 \cdot D} \right) \cdot \langle V_{IN} \rangle_T \tag{7.18}$$

In order to define the difference between the AC and DC operation, the gain factor G is introduced:

$$G = \frac{\langle V_0 \rangle_T}{\langle V_{IN} \rangle_T} = = \frac{1-D}{1-2 \cdot D} \tag{7.19}$$

Taking into account the value of gain factor G, it is possible to derive the expression of the output voltage in the case of AC operation:

$$V_{AC} = M \cdot G \cdot V_{IN}, \tag{7.20}$$

where M is a modulation index that varies from -1 up to 1 and is defined by the control system.

Fig. 7.10A shows a diagram that illustrates a voltage stress across capacitors and output ZS network voltage in the case of AC operation and equally distributed ST states. The voltage values are presented in the relative units where one voltage unit corresponds to amplitude of the nominal phase voltage.

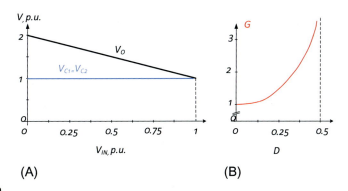

FIG. 7.10

Voltage across capacitors and ZS network as a function of input voltage in the case of AC application (A) and gain factor as a function of ST duty cycle (B).

230 CHAPTER 7 Resonant and Z-source multilevel inverters

It should be noted that in the buck mode, a converter works like a traditional VSI and the ZS network is not involved in its operation. It can be seen that the lowest input voltage corresponds to the highest voltage across the ZS network. This voltage is very important because it defines the voltage across semiconductors. The gain factor is demonstrated in Fig. 7.10B. In the case where the ST duty cycle tends to 0.5, the gain tends to infinity. In a real system the gain is limited by losses in the converter.

The main conclusion from Fig. 7.10 is that, in order to provide the wide input voltage range regulation, all components have to be rated for significant voltage stress.

Fig. 7.11A shows a diagram that illustrates a voltage stress across capacitors and output ZS network voltage in the case of DC operation and equally distributed ST states. Fig. 7.11B shows the boost factor.

Contrary to AC operation, the lowest input voltage corresponds to the lowest voltage stress across the capacitors, while the voltage across the ZS network remains constant and is equal to the nominal DC-link voltage.

Taking into account Eqs (7.8) and (7.10)–(7.12) and the power balance for a qZS network expressed in Eq. (7.15), the expressions for voltage stress across capacitors can be derived:

$$\langle V_{C1}^{C1}\rangle_T = \left(\frac{1-D}{1-2\cdot D}\right)\cdot \langle V_{IN}\rangle_T, \qquad (7.21)$$

$$\langle V_{C2}^{C2}\rangle_T = \left(\frac{D}{1-2\cdot D}\right)\cdot \langle V_{IN}\rangle_T. \qquad (7.22)$$

The gain and boost factors are the same as for ZS networks.

The voltage stress across capacitors as a function of input voltage in relative units is illustrated in Fig. 7.12. Fig. 7.12A shows AC operation, and Fig. 7.12B shows DC

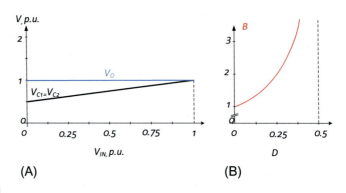

(A)   (B)

FIG. 7.11

Voltage across capacitors and ZS network as a function of input voltage in the case of DC application (A), and boost factor as a function of ST duty cycle (B).

7.2 General operating principle of impedance-source networks

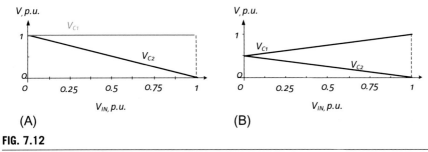

FIG. 7.12

Voltage across capacitors and qZS network as a function of input voltage in the case of AC application (A) and DC application (B).

operation. It can be seen that dependencies are similar to the ZS network. The key difference lies in the unequal voltage distribution across capacitors. The capacitor $C2$ in the qZS network has significantly lower voltage stress. The sum of the voltage across the capacitors is constant and is equal to the reference DC-link level.

From the preceding analysis it can be seen that rising gain or boost factors that correspond to the input voltage decreasing lead to voltage stress across capacitors and semiconductors in the AC mode. In order to eliminate this phenomena, other different IS networks are proposed. Most of them are proposed for increasing gain factors, which in turn leads to the ST states shortcutting and increasing of the effective voltage applied to the output.

Table 7.3 summarizes representatives of IS networks in terms of boost factor B, voltage across capacitors relative to the input voltage V_{IN}, and peak output voltage V_O relative to input voltage V_{IN}.

where n, $N1$, $N2$, $N3$ is turns ratio or numbers that correspond to Fig. 7.7. Due to the presence of the coupled inductors or transformers, the boost and gain factor can

Table 7.3 Basic properties and expressions of impedance source networks.

Name	Boost factor (B)	V_{C1}^*	V_{C2}^*
Z-source	$\frac{1}{1-2 \cdot D}$	$\frac{1-D}{1-2 \cdot D}$	$\frac{1-D}{1-2 \cdot D}$
Quasi-Z-source	$\frac{1}{1-2 \cdot D}$	$\frac{1-D}{1-2 \cdot D}$	$\frac{D}{1-2 \cdot D}$
Quasi-T-source	$\frac{1}{1-\frac{N_1}{N_3} \cdot D}$	$\frac{1-D}{1-\frac{N_1}{N_3} \cdot D}$	$\frac{1-D}{1-\frac{N_1}{N_3} \cdot D}$
T(trans)-quasi-Z-source with CCM	$\frac{1}{1-(n+1) \cdot D}$	$\frac{1-D}{1-(n+1) \cdot D}$	$\frac{n \cdot D}{1-(n+1) \cdot D}$
LCCT quasi-Z-source A-source	$\frac{1}{1-\left(2+\frac{N_2}{N_1}\right) \cdot D}$	$\frac{1-D}{1-\left(2+\frac{N_2}{N_1}\right) \cdot D}$	$\frac{\left(1+\frac{N_2}{N_1}\right) \cdot D}{1-\left(2+\frac{N_2}{N_1}\right) \cdot D}$

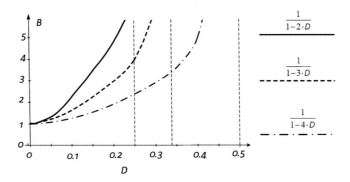

FIG. 7.13

The boost factor dependencies as a function of the duration of the ST duty cycle.

be increased. Fig. 7.13 illustrates the boost factor dependencies on the ST duty cycle for different IS network representatives in the following conditions:

$$n = \frac{N_1}{N_3} = \frac{N_2}{N_1} = 2. \qquad (7.23)$$

It can be seen that the turns ratio increasing leads to a significant boost factor increase, which in turn leads to the ST duty cycle decreasing, voltage spike decreasing, and conduction losses decreasing. At the same time, another analysis [25] revealed that such solutions based on a coupled inductor or transformer have extra volume and cost.

Also, it was shown that in a very general case, in order to provide higher boost of input voltage with the same ripples, much larger passive components are required for any IS network solution. Also, an interesting conclusion is that the overall size of the similar IS-based converters designed for identical operating conditions is the same.

7.3 Overview of multilevel inverters

A modern trend in power electronics involves modular and multilevel converter applications. Industry and academia are both paying increased attention to multilevel converters as one of the preferred choices for power electronic conversion for high-power applications, due to the fact that they can achieve high power using mature medium-power semiconductor technology. Several review papers have presented good classifications and descriptions of the multilevel converters [56–59]; for instance, many types of multilevel inverters (MLIs) are covered in [56]. Fig. 7.14 shows a generalized simplified classification of MLIs [27].

MLIs have advantages over the conventional and very well-known two-level inverters. These advantages include improved output quality and larger nominal power in the converter.

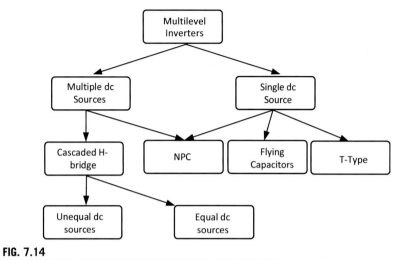

FIG. 7.14

Classification of the multilevel inverters.

Multilevel converters are a good solution for low-power and low-voltage applications as well. Their reduced voltage stress allows the use of fast MOSFET semiconductors among industrially verified Si technologies. In particular, three-level (3 L) inverters have attracted increasing attention in industrial applications, such as in motor drives [60], active filters [61, 62], and renewable energy systems [63]. The result is higher power quality, better electromagnetic compatibility, lower switching losses, and no need for a transformer at the distribution voltage level [64–66].

Based on [56], it can be concluded that the neutral-point-clamped (NPC) inverter is the most attractive solution for industrial applications. Fig. 7.15 shows the main representatives of MLIs with a single-input voltage source. A conventional NPC is illustrated in Fig. 7.15A, while Fig. 7.15B shows an active NPC (ANPC) in which diodes are replaced by active switches, which is justified by a reduction in conduction losses.

A T-type inverter is shown in Fig. 7.15C. It has a reduced number of semiconductors, but switches S_1 and S_2 have full DC-link voltage stress.

The flying capacitor (FC) inverter is quite a popular solution in industrial applications [67]. It has a reduced number of semiconductors compared to the NPC inverter. Fig. 7.15D shows a 3 L version of the FC inverter.

The NPC inverter has become popular because of its lower number of capacitors, particularly in the 3 L case. However, the NPC structure can be extended to a higher number of levels and phases, as shown in Fig. 7.16.

Fig. 7.16A shows a three-phase extension of the NPC inverter, while Fig. 7.16B shows the extension of the NPC to a 5-level solution, which in a very general case can be extended to any N-level solution.

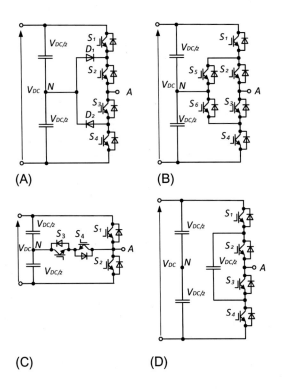

FIG. 7.15

Main types of MLIs: (A) conventional NPC, (B) active NPC, (C) T-type inverter, (D) flying capacitor inverter.

One of the drawbacks of the NPC inverter most frequently analyzed is the neutral point control, or capacitor voltage balance. Among other characteristics, this depends on the modulation index, dynamic behavior, and load conditions, which can produce a voltage difference between both capacitors, shifting the neutral point and causing undesirable distortion at the converter output [56]. This limits the number of levels in practical applications.

The cascaded H-bridge (CHB) solution (Fig. 7.17) is well suited for high-power and high-level applications because of the modular structure that enables higher voltage operation with classic low-voltage semiconductors. The phase shifting of the carrier signals moves the frequency harmonics to the higher frequency side, and this, together with the high number of levels, enables a reduction of the average device switching frequency and lower losses. However, it requires a large number of isolated DC sources, which, for instance, could be realized in PV panels.

At the same time, a great deal of research is being dedicated to further improvement of the MLI. Fig. 7.18 shows the solution based on the cascaded approach, with a reduced number of switches [68]. It should be noted that the described solutions are

7.3 Overview of multilevel inverters

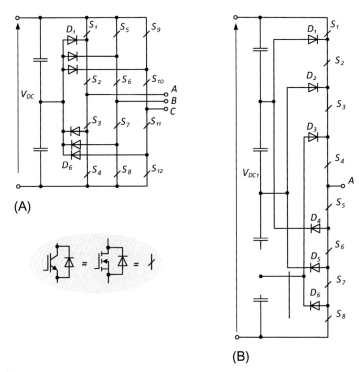

FIG. 7.16

Three-phase 3 L NPC inverter (A) and single-phase five-level NPC inverter (B).

basic examples. The principle of cascading can be applied to any MLI solution, and the number of levels is not limited.

Finally, as discussed in the following sections, the MLI concept can be applied in impedance-source inverters.

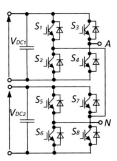

FIG. 7.17

Single-phase cascaded H-bridge inverter.

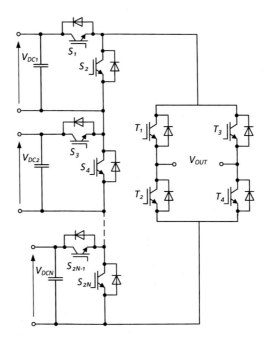

FIG. 7.18

Single-phase cascaded multilevel inverter with reduced switch count.

7.4 Resonant multilevel inverters: Main circuits

So far, multilevel converters have been studied and operated in the hard-switching mode. Soft-switching methods can be applied by exploiting the flexibility of the inherently cellular topology [69].

Soft-switching conditions can be achieved using two approaches. First, some special loads can themselves provide the conditions necessary for ZVS: the converter topology remains unchanged. Second, auxiliary circuits can be added to modify the current and voltage waveforms of the switches to allow for the ZVS conditions with any type of load.

In the first case, the solution obviously depends on the load impedance. Another solution is to control the current ripple by adding an output filter to achieve ZVS, whatever the load impedance [70, 71].

The second approach involves modifying the converter topology to guarantee the soft-switching conditions for each cell. ZVS conditions are ensured by the auxiliary inductor current ripple in the commutation cell and not by use of ripple on the load current. An example of such an approach is shown in Fig. 7.19.

In order to connect auxiliary networks and still respect the symmetry of the commutation cell, a voltage midpoint is created within each flying capacitor.

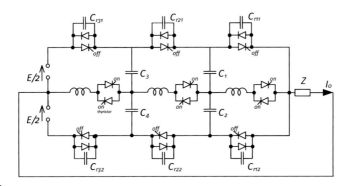

FIG. 7.19

Topology of multicell converter with an auxiliary resonant commutated pole [69].

The stability condition for these voltage dividers is defined by stating that the average current in the auxiliary inductors must be zero. In steady state, the average inductor voltages are zero: all the inductor terminals are at the same average potential. For each flying-capacitor pair, the sum of the two capacitor voltages must stay fixed and the difference between the two is modulated as a function of the duty cycle. Several passive resonant circuit variations discussed in previous sections are possible.

This solution offers numerous advantages over its counterparts. It reduces total converter losses, it does not increase the electrical stress on the main switches, and the auxiliary switches only operate during the main switches' commutations. In addition, this technique requires no alteration of the classical PWM control modes, and the auxiliary circuit can tolerate load variations.

The further extension of the resonant multilevel inverters with auxiliary switches is shown in Fig. 7.20. This demonstrates how the auxiliary resonant commutated poles (ARCPs) can be integrated to any multilevel inverter [72]. To prevent turn-on of the main switches at higher voltages, an accurate zero-voltage (ZV) detection is required for these soft-switching inverters. The turn-on times of the main and the auxiliary switches have to be synchronized.

These solutions belong to the 3 L NPC inverters [73–75]. The operating principle is very similar, the difference being the number of auxiliary switches and passive resonant elements. Fig. 7.21 shows the ARCP application for the flying capacitor solution. The ARCP circuit is identical and the modular structure can be maintained [69, 76].

It should be noted that all these circuits were proposed for high-power solutions when the benefits from soft switching overcome the cost of auxiliary circuit implementation. Also, the demonstrated ARCP can be implemented in any other multilevel structure with any N levels.

However, recent developments in high-voltage wide-bandgap semiconductors with significantly better dynamic performance are leading to loss of interest in this topic.

238 CHAPTER 7 Resonant and Z-source multilevel inverters

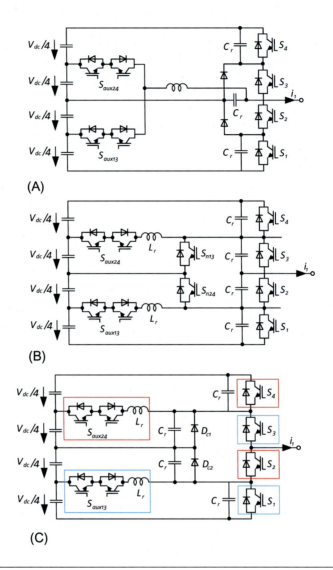

FIG. 7.20

Multilevel NPC topologies with auxiliary resonant poles: 3 L NPC ARCP inverters (A) and (B), 3 L NPC ARCP topology with related switches (C).

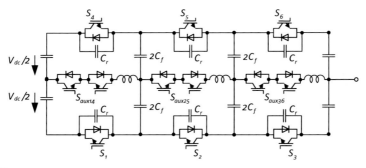

FIG. 7.21

Four-level flying capacitor ARCP inverter.

7.5 Impedance source-derived multilevel inverters: Control, benefits, and applications

The benefits of ZSI and its derived network include the reduction of the energy conversion stages in some specific applications and that the advantages of multilevel bridges include reduced voltage stress, which allows use of fast semiconductors and smaller size of output filters, among others. Because of this, recent solutions based on the combination of both of them are presented in this section.

Fig. 7.22 shows different three-level three-phase NPC topologies based on IS networks with discontinuous conduction mode (DCM) of the input current. The first single-stage buck-boost MLI (Fig. 7.22A) was proposed in [77] as the logical extension of the two-level inverter and ZSI. This configuration uses two IS networks for boosting their input voltage to a higher DC-link voltage. It is easy to observe that this is not the best solution from an economical point of view, since it uses two isolated input voltage sources and a number of passive elements, which can increase the cost, size, and weight of the inverter. To decrease the number of passive components, the Z-source NPC inverter with a single IS network was proposed in [78]. However, this topology must also be supplied from two input voltage sources (Fig. 7.22B). In those topologies, the voltage boost is expressed as

$$V_{dc} = \frac{1}{1-2D_S} V_{IN} \tag{7.24}$$

where V_{IN} is a DC input voltage.

By the introduction of the high-frequency transformer and two additional capacitors, the Z-source NPC inverter with a single IS network could be supplied from a single input voltage source (Fig. 7.22C) [79–82]. Fig. 7.7D shows a similar solution with a double transformer and separated input voltage DC sources. The main difference lies in the boost expression

$$V_{dc} = \frac{1}{1-(1+n)D_S} V_{IN}. \tag{7.25}$$

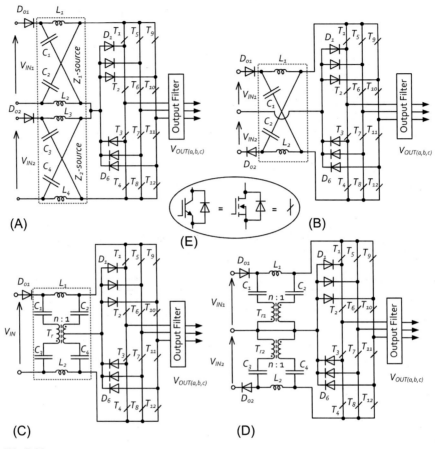

FIG. 7.22

3 L three-phase NPC topologies based on different IS networks. (A) With two and separate Z-source network and input voltage sources, (B) With single IS network and separate input voltage source, (C) Transformer Z-source NPC with single input voltage source, (D) Transformer Z-source NPC inverter with separate input voltage sources, (E) Generalization of power switch.

Using a transformer with n different from 1:1, an input voltage gain higher than that of the traditional IS network can be achieved.

Fig. 7.23A shows a buck/boost four-level inverter. In this case, each input power source is distributed among several IS networks. The idea was presented in [79] and it is an example of extension of any IS network to MLI based on the diode clamped configuration. Because of the possibility of separately regulating the output voltage in each Z-source, such MLIs are suitable to be applied in supply systems with locally

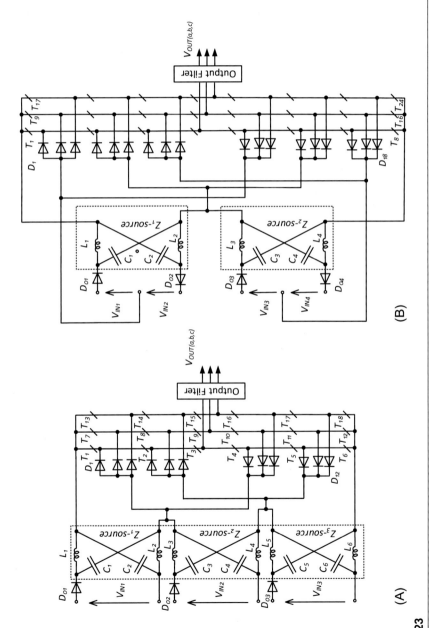

FIG. 7.23

(A) Four-level Z-source inverter of diode clamped type. (B) Five-level Z-source inverter of diode clamped type.

dispersed energy sources. Fig. 7.23B) illustrates multiple DC-link sources as further extension of the idea of the multilevel diode clamped topology [83] including just two IS networks.

Fig. 7.9 shows several modifications of the previous buck/boost MLI. Fig. 7.24A) shows the three-phase three-level DC-link cascaded (DCLC) inverter [84], where the passive elements are the same as in the NPC with two IS networks. The only difference between the 3 L NPC inverter and the DCLC inverter is in the asymmetrical blocking voltage in the transistors and the absence of the clamping diodes. A variation based on a dual configuration is also presented [84] in Fig. 7.24B and C, with two and single IS networks, respectively. Their DC-link voltage can be two times lower in the case of n equal to 2. The main difference between the dual solution with separated and single IS networks is the power flow. In the second case, the number of passive components is half, but their size is larger.

Cascaded solutions based on the Z-source networks have already been presented [85, 86]. Simple cascading is described in [86]. Two ZSNs, two isolated input voltage sources, and two two-level full-bridge (FB) inverters provide a five-level output voltage per single phase.

Fig. 7.25 shows the general structure of the most complex cascading of the hybrid-sourced network in the 3 L NPC [85]. A seven-level Z-source based inverter is carried out by means of cascading traditional, embedded, and DC-link embedded voltage-type Z-source that were proposed in [86]. This configuration requires three isolated input voltage sources and N networks. The total boost of this converter is:

$$V_{dc} = \frac{1}{1-(1+N)D_S} V_{IN} \qquad (7.26)$$

where $V_{IN} = V_{IN}1 + V_{IN}2 + V_{IN}3$.

This method requires that the number of N cascaded networks must always be odd with the middle network notated as

$$K = \frac{N+1}{2}. \qquad (7.27)$$

Another proposed cascaded configuration in [87] is based on $N - 1$ additional capacitors and $2(N - 1)$ additional diodes for connecting N IS networks together at their respective DC links. This so-called DC-link cascaded solution achieves very high boost performance:

$$V_{dc} = \frac{1}{(1-2D_S)^{\frac{N+1}{2}}} V_{IN} \qquad (7.28)$$

Paper [88] presents the description of a new inverter topology based on a mixture of cascaded basic units and one FB (Fig. 7.26). The basic unit includes one IS network, one input DC voltage source and two switches, generating two voltage levels. The basic unit can operate in three modes: zero, active, and ST states. In the ST state, both switches S1 and S2 are conducting and the output voltage is zero. The active state is generating when only S1 is conducting. The zero state corresponds to the S2

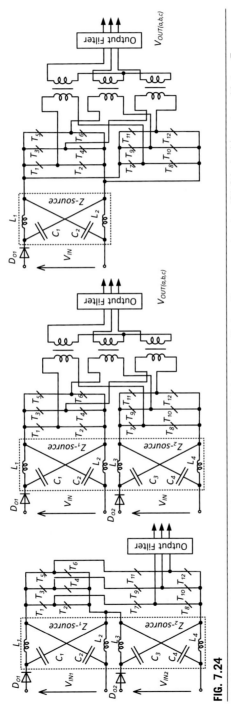

FIG. 7.24

Modifications of the 3 L Z-source based inverters: (A) 3 L DCLC inverter with two Z-source networks, (B) 3 L dual inverter with two Z-source networks, (C) 3 L dual inverter with single Z-source network.

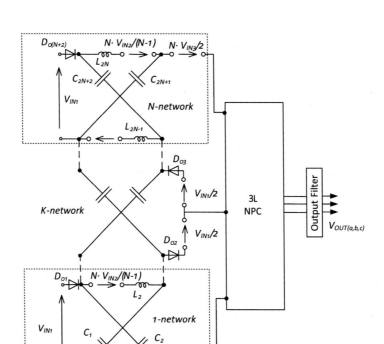

FIG. 7.25

Seven-level Z-source based inverter.

conduction. In this solution, the overall number of power semiconductor switches is reduced with respect to the traditional MLIs. This configuration allows that any N-level topology is achievable due to the described principle and also can be extended to three-phase systems.

Further development of the IS network-based buck-boost MLIs is carried out by involving trans-ZS and trans-qZS inverters. In particular, Fig. 7.27 shows two 3 L NPC solutions [89]. The circuit configuration of the trans-Z-source NPC inverter is shown in Fig. 7.27A and B shows the circuit configuration with the trans-quasi-Z-source NPC inverter, where the only difference is the location of the input voltage source. Both of them consist of a transformer to replace the two inductors in the original IS network and removing one capacitor from it. This can enhance the boosting capability of the IS network and reduce one passive component, with consequent lower size and cost of the system.

The main difference between the first and second approaches consists of the input current waveform, which corresponds to a CCM in the second case.

Another trans-Z-source NPC inverter was proposed in [90] and depicted in Fig. 7.28A. Fig. 7.28B shows the Γ-source inverter that was proposed in the same

7.5 Impedance source-derived multilevel inverters

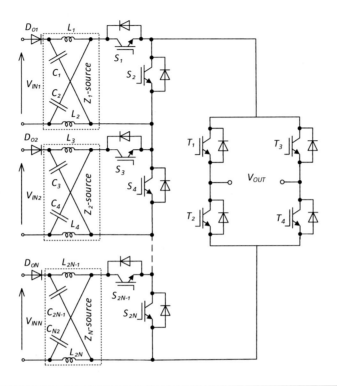

FIG. 7.26

Z-source-based MLI with reduction of switches.

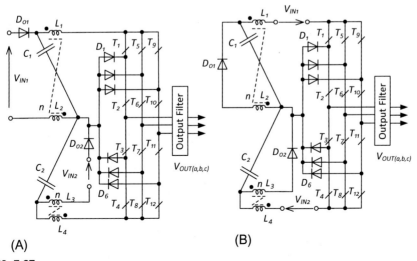

FIG. 7.27

(A) 3 L NPC trans-Z-source inverter and (B) 3 L NPC trans-quasi-Z-source inverter.

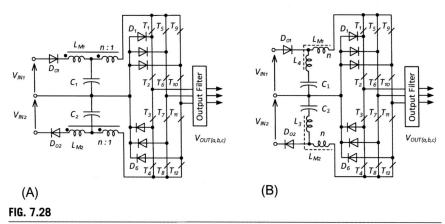

FIG. 7.28

(A) Other 3 L NPC trans-Z-source inverter and (B) 3 L NPC - Γ source inverter.

paper. Input voltage boost for the first topology is equal to Eq. (7.25), but for the second one:

$$V_{dc} = \frac{1}{1 - \frac{n}{n-1}D_S} V_{IN}. \qquad (7.29)$$

When an appropriate modulation technique is being designed for this kind of power converter, the following common requirements are set to carefully embed the ST states into the conventional switching signals:

- no violation of normalized volt-second balance in the output voltages during the fundamental period
- minimum number of extra commutations in the switches
- lower semiconductor stress
- minimum total harmonic distortion
- maximum voltage boost
- minimum complexity of implementation.

The initial work that deals with the development of switching signal generations for this family of inverters is found in [77]. The first PWM approach is derived from the optimized PWM sequences of a conventional two-level ZSI and the nearest three vectors (NTVs) modulation principle in a conventional NPC inverter. The arrangement of reference and carrier signals is depicted in Fig. 7.29A and, as a main feature, the resulting switching pattern avoids extra device switching signals (six per half switching period). Each impedance network (upper and lower) is short-circuited sequentially with equal time intervals (half the total shoot-through duration) in order to avoid DC voltage imbalance. In this strategy, as a core idea, the maximum and minimum voltage reference signals (in each switching period) are modified with an offset (MR1 and MR2) and then compared with the set of carriers to generate the upper and lower ST states (just one phase-leg, A, B, or C is shoot-through).

7.5 Impedance source-derived multilevel inverters

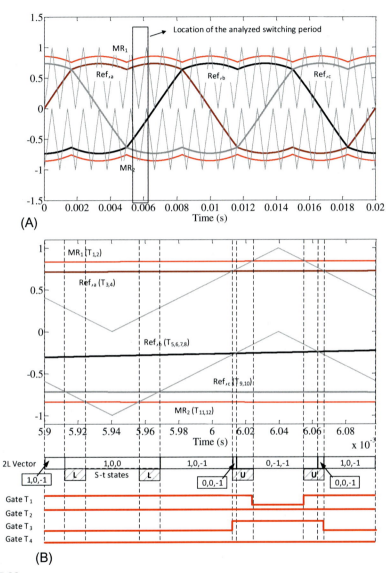

FIG. 7.29

NTV/MR PWM scheme. (A) Reference and carrier arrangement. (B) Vectorial analysis of a switching period.

Operating in this way, the volt-second average per switching cycle is maintained and a minimum harmonic distortion is obtained. To enhance the boost factor, the injection of third harmonic offset is also added to the reference signals.

Fig. 7.29B shows a vectorial analysis when the two-level effective voltage reference phasor is located in a certain triangle of the SV representation and the shoot-through interval (T0/2) becomes longer than the active state $\{0,0,-1\}$. It is interesting to remark that the intermediate located reference signal (Ref$_b$ in the illustrated case) does not trigger any shoot-through; hence this is a difference with the two-level Z-source PWM formulation. This PWM scheme is also valid for a trans-Z-source NPC inverter.

The second proposed principle from the aforementioned references (Fig. 7.30) is devoted to eliminating the common-mode voltage (if the reference grounding point is chosen between both IS networks). This switching pattern is created by a proper logical mapping between the two-level Z-source sequence and the seven states of the reduced common-mode (RCM) three-level vector diagram. Both the upper and lower Z-source networks are short-circuited simultaneously, since the dead time delays are not needed. The arrangement of reference and carrier signals is depicted in Fig. 7.30A. Fig. 7.30B shows the vectorial analysis.

In references [91, 92], the aforementioned concepts are extended and clarified, proposing a number of continuous and discontinuous PWM schemes for the same topology. Continuous edge insertion (EI) PWM with a symmetrical voltage boost is very similar to the NTV-derived technique, but involves getting fixed ST state positions into zero states and equal durations regardless of the sextant which the reference voltage is in. This fact minimizes the current ripple flowing across the Z-source network but at the expense of eight device commutations.

The so-called continuous modified reference (MR) PWM is the same as the first presented (NTV, Fig. 7.29), but it is much better clarified how it is necessary to shift some active states (in some particular positions) to compensate the zero output voltage generated by the ST states, keeping the normalized volt-second average. In order to reduce the number of state transitions, a new ST modulation strategy was derived based on the conventional 60° discontinuous PWM and the MR PWM. The modification included over MR PWM is just in the reference offsets, because these have to be changed periodically (which increases the complexity). Also, it is reported to be problematic that just one Z-source network per 60° sextant is short-circuited (which limits its application). The same reference signals but with triple offset are proposed in [93], and with the equal alternation of reference offsets to generate the ST, the common mode voltage is limited. Discontinuous schemes are not recommended since they give rise to a large current ripple. If the optimization of the control algorithm is the priority, a good solution is found in [94], where the SVPWM and triangular-comparison PWM approaches are combined for controlling the converter operation.

The APOD with 180° arrangement (Fig. 7.31A and B) shows the arrangement of references and carriers and the vectorial analysis, respectively; it is considered for carriers because it creates multiple null intervals for ST insertion, which are

7.5 Impedance source-derived multilevel inverters

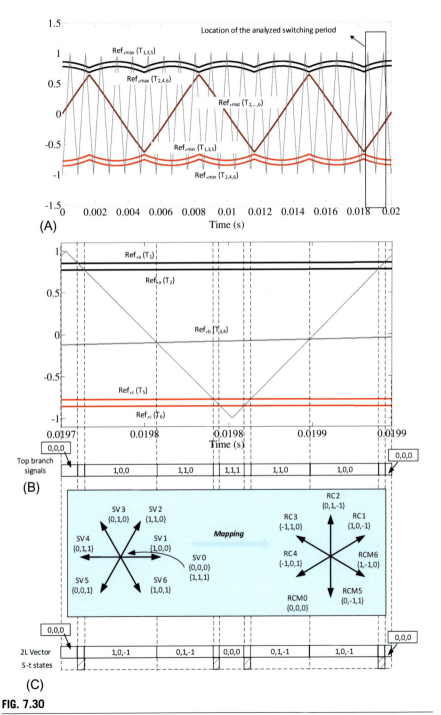

FIG. 7.30

RCM PWM scheme. (A) Reference and carrier arrangement for two-level (2 L) formulation. (B) Vectorial analysis for 2 L formulation in a switching period. (C) Logic mapping from 2 L to RCM 3 L.

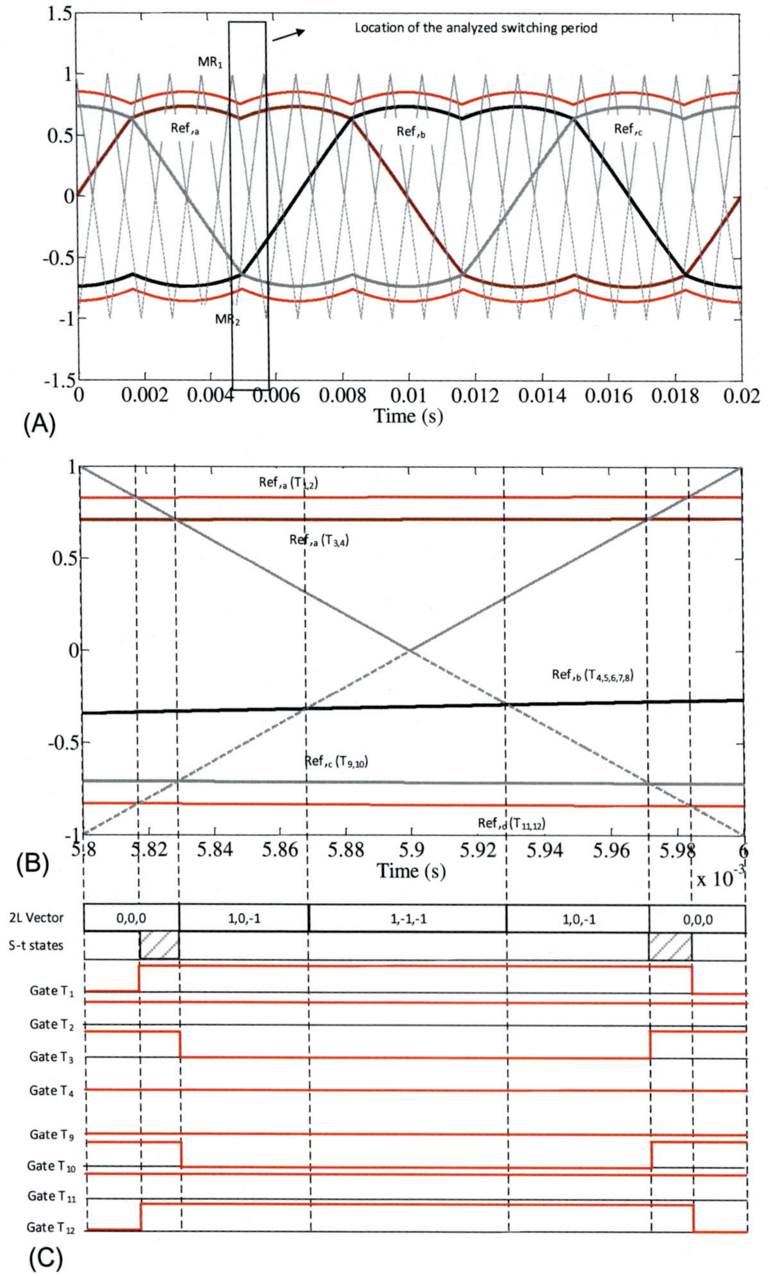

250 CHAPTER 7 Resonant and *Z*-source multilevel inverters

FIG. 7.31

APOD scheme. (A) Reference and carrier arrangement. (B) Vectorial analysis of a switching period.

generated with the aim of looking for the minimum losses. ST is achieved by a proper synchronization of turning on switches from two selected phase legs, instead of turning on all switches from the same phase leg simultaneously.

Additional references can be programmed with the offset (T_0/T_{sw}) to the maximum and minimum reference in each switching period. As this technique is not based on NTV, it has higher THD_U values. To resolve this undesired result, the application of PD technique is studied in [95] (this topology is called reduced element count (REC)). A similar approach to this PWM based on NTV is detailed in Ref. [96] and derived from SVPWM for a simpler implementation and improved harmonic performance [97]. Basically, duty ratios of the nearest three vectors to be applied by the converter in any SVPWM triangle are calculated. After this, a new methodology is proposed, which generates a new switching sequence in comparison with the previous one, to insert the ST states while the volt-second balance and minimal commutation count are maintained.

In [98], a new phase-shifted PWM devoted to qZ source CMC is proposed. Its scheme is composed of two sets of three reference signals (for the left and right leg, respectively) and one set of three carriers, one per module (each layer of the modules has the same carrier). The most particular feature of this method is that the carriers change their amplitudes according to the envelopes of the references. To embed the ST states, the SBC is included. As the main achievement, a power loss reduction is remarked. In Ref. [99] a SVPWM for a single-phase qZS-CMC can be found, where the total ST duration is equally distributed into the zero states. The extension of this SVM-derived technique to the three-phase system is explained in Refs. [100, 101].

7.6 Conclusions

The main outcome of this chapter is that resonant multilevel inverters and multilevel impedance source inverters have correspondent niches in power electronics.

The resonant multilevel inverters are well-known solutions and are used in high-power electronics, mostly for medium-voltage applications. Resonant circuits allow minimizing switching losses and as a result switching frequency can be increased.

The impedance-source multilevel inverters have attracted significant research interest, and there are many research papers devoted to possible derivations. A variety of impedance-source networks along with a variety of multilevel inverters multiplies the number of derived impedance-source multilevel inverters.

In most cases, the most significant advantage in the voltage splitting is derived from the multilevel nature of the converters. It was also demonstrated that the difference lies in the input current waveform and voltage distribution across the capacitors, but the overall size of the converters designed for identical operating conditions is the same. A wide input voltage regulation requires larger passive elements. Also, the voltage stress across semiconductors is increasing. The modulation techniques with unequally distributed ST states decrease the voltage stress across the semiconductors. In order to mitigate such oscillations, passive element size should be increased as well. At the same time, despite the great number of research papers, there are no examples of industrial applications.

References

[1] P.R.K. Chetty, Resonant power supplies: their history and status, IEEE Aerospace and Electronic Systems Magazine 7 (4) (1992) 23–29.

[2] G. Hua, F.C. Lee, Soft-switching techniques in PWM converters, IEEE Trans. Ind. Electron. 42 (6) (Dec. 1995) 595–603.

[3] R. Lin, C. Lin, "Design Criteria for Resonant Tank of LLC DC-DC Resonant Converter," IECON 2010 - 36th Annual Conference on IEEE Industrial Electronics Society, Glendale, AZ, 2010, pp. 427–432.

[4] M. Salem, et al., Three-phase series resonant DC-DC boost converter with double LLC resonant tanks and variable frequency control, IEEE Access 8 (2020) 22386–22399.

[5] S.A. Chickamenahalli, J.J. Cathey, A resonant-commutated-link variable-frequency converter, IEEE Trans. Ind. Electron. 45 (2) (1998) 207–216.

[6] S. Liu, F. Zhang, State variable derivation with numerical approach and efficiency optimisation method for phase-shift LLC converters under wide voltage-gain range, IET Power Electron. 12 (7) (2019) 1752–1762.

[7] M. Abbasi, J. Lam, A modular SiC-based step-up converter with soft-switching-assisted networks and internally coupled high-voltage-gain modules for wind energy system with a medium-voltage DC-grid, IEEE Journal of Emerging and Selected Topics in Power Electronics 7 (2) (June 2019) 798–810.

[8] D. Maksimovic, S. Cuk, A general approach to synthesis and analysis of quasi-resonant converters, IEEE Trans. Power Electron. 6 (1) (Jan. 1991) 127–140.

[9] Y. Noh, M. Oh, M. Ryu, J. Kim, C. Won, Isolated Bi-directional DC/DC converter using quasi-resonant ZVS, in: *2014 IEEE International Conference on Industrial Technology (ICIT)*, Busan, 2014, pp. 514–518.

[10] S.I. Annie, K.M. Salim, Z. Tasneem, M.R. Uddin, Design and performance analysis of a ZVS parallel quasi resonant converter for a solar based induction cooking system, in: 2016 IEEE Region 10 Conference (TENCON), Singapore, 2016, pp. 2638–2641.

[11] S. Xu, W. Shen, Q. Qian, J. Zhu, W. Sun, H. Li, An efficiency optimization method for a high frequency quasi-ZVS controlled resonant flyback converter, in: 2019 IEEE Applied Power Electronics Conference and Exposition (APEC), Anaheim, CA, USA, 2019, pp. 2957–2961.

[12] U. Mumtahina, P. Wolfs, A comparison study between series resonant and zero-voltage-resonant-transition DC-DC converters, in: 2015 Australasian Universities Power Engineering Conference (AUPEC), Wollongong, NSW, 2015, pp. 1–6.

[13] A. Kar, M. Sengupta, B. Barman, Design, fabrication, simulation and testing of a Phase Modulated Resonant Transition Converter, in: 2014 IEEE International Conference on Power Electronics, Drives and Energy Systems (PEDES), Mumbai, 2014, pp. 1–6.

[14] M.L. Martins, J. Russi, H.L. Hey, Low reactive energy ZCZVT PWM converters: Synthesis, analysis and comparison, in: 2005 IEEE 36th Power Electronics Specialists Conference, Recife, 2005, pp. 1234–1240.

[15] W. Chen, X. Wu, L. Yao, W. Jiang, R. Hu, A step-up resonant converter for grid-connected renewable energy sources, IEEE Trans. Power Electron. 30 (6) (2015) 3017–3029.

[16] R.N. Beres, X. Wang, F. Blaabjerg, M. Liserre, C.L. Bak, Optimal design of high-order passive-damped filters for grid-connected applications, IEEE Trans. Power Electron. 31 (3) (2016) 2083–2098.

[17] H. Sarnago, Ó. Lucía, A. Mediano, J.M. Burdío, Multi-MOSFET-based series resonant inverter for improved efficiency and power density induction heating applications, IEEE Trans. Power Electron. 29 (8) (2014) 4301–4312.

[18] H. Sarnago, Ó. Lucía, A. Mediano, J.M. Burdío, Design and implementation of a high-efficiency multiple-output resonant converter for induction heating applications featuring wide bandgap devices, IEEE Trans. Power Electron. 29 (5) (2014) 2539–2549.

[19] M.T. Outeiro, G. Buja, D. Czarkowski, Resonant power converters: an overview with multiple elements in the resonant tank network, IEEE Industrial Electronics Magazine 10 (2) (2016) 21–45.

[20] F.Z. Peng, Z-source inverter, IEEE Trans. Ind. Applicat. 39 (2) (2003) 504–510.

[21] J. Anderson, F.Z. Peng, Four quasi-Z-Source inverter, in: Proceeding of the 2008 IEEE Power Electronics Specialists Conference (PESC'08), 2008, pp. 2743–2749.

[22] F.Z. Peng, A. Joseph, J. Wang, M. Shen, L. Chen, Z. Pan, E.O. Rivera, Y. Huang, Z-source inverter for motor drives, IEEE Trans. Power Electron. 20 (2005) 857–863.

[23] F.Z. Peng, X. Yuan, X. Fang, Z. Qian, Z-source inverter for adjustable speed drives, IEEE Trans. Power Electron. 1 (2) (2003) 33–35.

[24] Y.P. Siwakoti, F. Peng, F. Blaabjerg, P.C. Loh, G.E. Town, Impedance source networks for electric power conversion part-I: a topological review, IEEE Trans. Power Electron. 30 (2) (2015) 699–716.

[25] O. Husev, T. Shults, D. Vinnikov, C. Roncero-Clemente, E. Romero-Cadaval, A. Chub, Comprehensive comparative analysis of impedance-source networks for DC and AC application, Electronics 8 (2019) 405.

[26] T. Shults, O. Husev, J. Zakis, Overview of impedance source networks for voltage source inverters, in: Proc. Micro/Nanotechnologies and Electron Devices (EDM), 2015 16th International Conference of Young Specialists, 2015, pp. 514–520.

[27] O. Husev, F. Blaabjerg, C. Roncero-Clemente, E. Romero-Cadaval, D. Vinnikov, Y.P. Siwakoti, R. Strzelecki, Comparison of impedance-source networks for two and multilevel Buck-boost inverter applications, IEEE Trans. Power Electron. 31 (2016) 7564–7579.

[28] B. Gadalla, E. Schaltz, Y. Siwakoti, F. Blaabjerg, Thermal performance and efficiency investigation of conventional boost, Z-source and Y-source converters, in: Proceeding of the 16th International Conference on Environment and Electrical Engineering (EEEIC), 2016, pp. 1–6.

[29] A. Battiston, J.P. Martin, E.D. Miliani, B. Nahid-Mobarakeh, S. Pierfederici, F. Meibody-Tabar, Comparison criteria for electric traction system using z-source/quasi z-source inverter and conventional architectures, IEEE J. Emerg. Sel. Topics Power Electron. 2 (3) (2014) 467–476.

[30] Y.P. Siwakoti, F. Blaabjerg, V.P. Galigekere, A. Ayachit, M.K. Kazimierczuk, Y-source impedance network, IEEE Trans. Power Electron. 31 (12) (2016) 8081–8087.

[31] W. Mo, P.C. Loh, F. Blaabjerg, Voltage Type Γ-Source Inverters with Continuous Input Current and Enhanced Voltage Boost Capability, in: Proceeding of the 15th International Power Electronics and Motion Control Conference, EPE-PEMC 2012 ECCE Europe, Novi Sad, Serbia, 2014, pp. 1–8.

[32] Y.P. Siwakoti, F. Blaabjerg, P.C. Loh, New magnetically coupled impedance (Z-) source networks, IEEE Trans. Power Electron. 31 (11) (2016) 7419–7435.
[33] R. Strzelecki, M. Adamowicz, B. Balkowski, N. Strzelecka, "The buck-boost inverter circuit especially designed for single-stage power conversion", PL Patent Application P386084, Sep. 09, 2008.
[34] R. Strzelecki, M. Adamowicz, N. Strzelecka, W. Bury, New type T-source inverter, in: Proceeding of the CPE 2009, 2009, pp. 191–195.
[35] R. Strzelecki, M. Adamowicz, B. Balkowski, N. Strzelecka, "Multi-level inverter circuit especially for voltage boost," PL Patent Application P386085, March. 29, 2010.
[36] T. Shults, O. Husev, F. Blaabjerg, Design and Comparison of Three-Level Three-Phase T-Source Inverters, in: Proceeding of International Conference POWERENG'2015, 2015, pp. 1–6. 11–13 May.
[37] W. Mo, P.C. Loh, F. Blaabjerg, P. Wang, Trans-Z-source and Γ-Z-source neutral-point-clamped inverters, IET Power Electron. 8 (3) (2015) 1–7.
[38] M. Adamowicz, J. Guzinski, D. Vinnikov, N. Strzelecka, Trans-Z-source-like inverter with built-in DC current blocking capacitors, in: In Proceeding of the CPE 2011, 2011, pp. 137–143, https://doi.org/10.1109/CPE.2011.5942221.
[39] M. Adamowicz, R. Strzelecki, F.Z. Peng, J. Guzinski, H. Abu-Rub, New type LCCT-Z-source inverters, in: Proceeding of the 14th European Conference on Power Electronics, 2011, pp. 1–10.
[40] M. Adamowicz, J. Guzinski, R. Strzelecki, F.Z. Peng, H. Abu-Rub, High Step-Up Continuous Input Current LCCT-Z-Source Inverters for Fuel Cells, in: Proceeding of the IEEE Energy Conversion Congress and Exposition, ECCE, September, 2011, pp. 2276–2282.
[41] M. Adamowicz, LCCT-Z-source inverters, in: Proceeding of the 10th International Conference on Environment and Electrical Engineering, EEEIC, 2011, May, 2011, pp. 1–6.
[42] J.-K. Park, Y.-S. Shin, Y.-G. Jung, Y.-C. Lim, LCCT Z-Source DC-DC Converter with the Bipolar Output Voltages for Improving the Voltage Stress and Ripple, The Transactions of the Korean Institute of Power Electronics 18 (2013) 91–102.
[43] T. Shults, O. Husev, F. Blaabjerg, J. Zakis, K. Khandakji, LCCT-derived three-level three-phase inverters, IET Power Electron. 10 (2017) 996–1002.
[44] Y.P. Siwakoti, F. Blaabjerg, V.P. Galigekere, M.K. Kazimierczuk, A-source impedance network, in: Proceeding of the IEEE Energy Conversion Congress and Exposition (ECCE), 2016, pp. 1–6.
[45] Y.P. Siwakoti, F. Blaabjerg, V.P. Galigekere, A. Ayachit, M.K. Kazimierczuk, A-source impedance network, IEEE Trans. Power Electron. 31 (12) (2016) 8081–8087.
[46] S. Rajakaruna, L. Jayawickrama, Steady-state analysis and designing impedance network of Z-source inverters, IEEE Trans. Ind. Electron. 57 (7) (2010) 2483–2491.
[47] Y.P. Siwakoti, F.Z. Peng, F. Blaabjerg, P.C. Loh, G.E. Town, S. Yang, Impedance-source networks for electric power conversion part II: review of control and modulation techniques, IEEE Trans. Power Electron. 30 (4) (2015) 1887–1905.
[48] Y. Liu, H. Abu-Rub, G. Ge, Z-source/quasi-Z-source inverters: Derived Networks, Modulations, Controls, and Emerging Applications to Photovoltaic Conversion, IEEE Ind. Electron. Magazine 8 (4) (2014) 32–44.
[49] M. Shen, Z-source inverter design, analysis, and its application in fuel cell vehicles, PhD thesis, Michigan State University, 2006.
[50] F.Z. Peng, M. Shen, Z. Qian, Maximum boost control of the Z-source inverter, IEEE Trans. Power Electron. 2 (4) (2005) 833–838.

[51] M. Shen, J. Wang, A. Joseph, F.Z. Peng, L.M. Tolbert, D.J. Adams, Constant boost control of the Z-source inverter to minimize current ripple and voltage stress, IEEE Trans. Industry Applic. 42 (3) (2006) 770–778.

[52] O. Ellabban, J. Van Mierlo, P. Lataire, Comparison between different PWM control methods for different Z-source inverter topologies, in: 13th IEE European Conference on Power Electronics and Applications, EPE '09, 8–10 September, 2009, pp. 1–11.

[53] Y. Huang, M.S. Shen, F.Z. Peng, J. Wang, Z-source inverter for residential photovoltaic systems, IEEE Trans. Power Electron. 21 (6) (2006) 1776–1782.

[54] R. Badin, Y. Huang, F.Z. Peng, H.G. Kim, Grid Interconnected Z-Source PV System, in: Proc. IEEE PESC'07, Orlando, FL, June, 2007, pp. 2328–2333.

[55] D. Sun, B. Ge, D. Bi, F.Z. Peng, Analysis and control of quasi-Z source inverter with battery for grid-connected PV system, Electr. Power Energy Syst. 46 (2013) 234–240.

[56] L.G. Franquelo, J. Rodriguez, J.I. Leon, S. Kouro, R. Portillo, M.A.M. Prats, The age of multilevel converters arrives, IEEE Ind. Electron. Mag. 2 (2) (Jun. 2008) 28–39.

[57] S. Kouro, M. Malinowski, K. Gopakumar, J. Pou, L.G. Franquelo, B. Wu, J. Rodriguez, M.A. Perez, J.I. Leon, Recent advances and industrial applications of multilevel converters, IEEE Trans. Ind. Electron. 57 (8) (2010) 2553–2580.

[58] J. Rodriguez, S. Bernet, B. Wu, J.O. Pontt, S. Kouro, Multilevel voltage-source-converter topologies for industrial medium-voltage drives, IEEE Trans. Ind. Electron. 54 (6) (2007) 2930–2945.

[59] J. Rodríguez, J.S. Lai, F.Z. Peng, Multilevel inverters: a survey of topologies, controls, and applications, IEEE Trans. Ind. Electron. 49 (4) (2002) 724–738.

[60] K.B. Lee, S.H. Huh, J.Y. Yoo, F. Blaabjerg, Performance improvement of DTC for induction motor-fed by three-level inverter with an uncertainty observer using RBFN, IEEE Trans. Energy Conv. 20 (2) (2005) 276–283.

[61] O. Vodyakho, C.C. Mi, Three-level inverter-based shunt active power filter in three-phase three-wire and four-wire systems, IEEE Trans. Power Electron. 24 (5) (2009) 1350–1363.

[62] H. Akagi, R. Kondo, A Transformerless hybrid active filter using a three-level Pulse-width modulation (PWM) converter for a medium-voltage motor drive, IEEE Trans. Power Electron. 25 (6) (2010) 1365–1374.

[63] C. Xia, X. Gu, T. Shi, Y. Yan, Neutral-point potential balancing of three-level inverters in direct-driven wind energy conversion system, IEEE Trans. Energy Conv. 26 (1) (2011) 18–29.

[64] M.I. Milanés-Montero, E. Romero-Cadaval, F. Barrero-González, Hybrid multiconverter conditioner topology for high-power applications, IEEE Trans. Ind. Electron. 58 (6) (2011) 2283–2292.

[65] J.C. Rosas-Caro, J.M. Ramirez, F.Z. Peng, A. Valderrabano, A DC–DC multilevel boost converter, IET PEL. 3 (1) (2008) 129–137.

[66] H.W. Ping, N.A. Rahim, J. Jamaludin, New three-phase multilevel inverter with shared power switches, Journal of Power Electronics 13 (5) (2013) 787–797.

[67] M.F. Escalante, J.C. Vannier, A. Arzande, Flying capacitor multilevel inverters and DTC motor drive applications, IEEE Trans. Ind. Electron. 49 (4) (2002) 809–815.

[68] M.R. Banaei, S.H. Hosseini, New cascaded multilevel inverter topology with minimum number of switches, Energ. Conver. Manage. 50 (11) (2009) 2761–2767.

[69] C. Turpin, L. Deprez, F. Forest, F. Richardeau, T.A. Meynard, A ZVS imbricated cell multilevel inverter with auxiliary resonant commutated poles, IEEE Trans. Power Electron. 17 (2) (2002).

[70] S. Boyer, H. Foch, J. Roux, M. Metz, Chopper PWM Inverter Using GTO's in Dual Thyristor Operation, in: Proceedings of European Power Electron Conference, 1987, pp. 383–389.

[71] H. Foch and J. Roux, "Static semi-conductor electrical energy converter apparatus," U.S. Patent 4 550 365, Oct. 29, 1985.

[72] P. Kollensperger, R.U. Lenke, S. Schroder, R.W. De Doncker, Design of a Flexible Control Platform for Soft-Switching Multilevel Inverters, IEEE Trans. Power Electron. 22 (5) (2007) 1778–1785.

[73] J.G. Cho, J.W. Baek, D.W.Y. Chung, Y. Won, Three-Level Auxiliary Resonant Commutated Pole ET Inverter for High Power Applications, in: IEEE-PESC Proceedings, 1996, pp. 1019–1026.

[74] R. Teichmann, S. Bernet, A multi-level ARCP voltage source converter topology, in: IEEE-IECON Proceedings, 1999, pp. 602–607.

[75] X. Yuan, G. Orglmeister, I. Barbi, ARCPI resonant snubber for the neutral-point-clamped inverter, IEEE Trans. Ind. Appl. 36 (2) (2000) 586–595.

[76] K. Sano, H. Fujita, Voltage-Balancing Circuit Based on a Resonant Switched-Capacitor Converter for Multilevel Inverters, IEEE Trans. Ind. Appl. 44 (6) (2008) 1768–1776.

[77] P.C. Loh, F. Gao, F. Blaabjerg, S.Y.C. Feng, K.N.J. Soon, Pulsewidth-modulated Z-source neutral-point-clamped inverter, IEEE Trans. on in Industry Applications 43 (5) (2007) 1295–1308.

[78] P.C. Loh, S.W. Lim, F. Gao, F. Blaabjerg, Three-level Z-source inverters using a single LC impedance network, IEEE Trans. Power Electron. 22 (2) (March 2007) 706–711.

[79] R. Strzelecki, M. Adamowicz, D. Wojciechowski, Buck-Boost Inverters with Symmetrical Passive Four-terminal Networks, in: In conference proceeding in Compatibility in Power Electronics, 2007. CPE '07, 2007, pp. 1–9.

[80] R. Strzelecki, W. Bury, M. Adamowicz, N. Strzelecka, New alternative passive networks to improve the range output voltage regulation of the PWM inverters, in: In conference proceeding in IEEE APEC, 2009, pp. 857–863.

[81] R. Strzelecki, D. Wojciechowski, M. Adamowicz, "Multilevel, multiphase inverter supplying by many sources, especially different voltage and non-connection sources". Patent Application P379977, Jun. 19, 2006.

[82] R. Strzelecki, D. Wojciechowski, M. Adamowicz, "Principle of symmetrization of the output voltage of the multilevel inverter supplying by many different voltage sources, especially with four-terminal impedance networks. PL Patent Application P379978, Jun. 19". 2006.

[83] F. Gao, P.C. Loh, F. Blaabjerg, R. Teodorescu, D.M. Vilathgamuwa, Five-level Z-source diode-clamped inverter, IET Power Electron. 3 (4) (2010) 500–510.

[84] P.C. Loh, F. Gao, F. Blaabjerg, Topological and modulation Design of Three-Level Z-source inverters, IEEE Trans. Power Electron. 23 (5) (2008) 2268–2277.

[85] D. Li, F. Gao, P.C. Loh, M. Zhu, F. Blaabjerg, Cascaded Impedance Networks for NPC Inverter, in: In conference proceeding in IPEC, 2010, pp. 1176–1180.

[86] A.S. Priyaa, R. Seyezhai, B.L. Mathur, "Design and Implementation of Cascaded Z-Source Multilevel Inverter". In conference proceeding in International Conference on Advances in Engineering, Science and Management (ICAESM), pp. 354–360.

[87] P.C. Loh, F. Gao, F. Blaabjerg, Embedded EZ-Source Inverter, IEEE Trans. Ind. Appl. 46 (1) (2010) 256–267.

[88] M.R. Banaei, A.R. Dehghanzadeh, E. Salary, H. Khounjahan, R. Alizadeh, Z-source-based multilevel inverter with reduction of switches, IET Power Electron. 5 (3) (2012) 385–392.

[89] W. Mo, P.C. Loh, D. Li, F. Blaabjerg, Trans-z-source neutral point clamped inverter, in: Conference proceeding in 6th IET International Conference on Power Electronics, Machines and Drives (PEMD 2012), pp. 1–5, 2012, 2012.

[90] W. Mo, P.C. Loh, F. Blaabjerg, P. Wang, Trans-Z-source and Γ-Z-source neutral-point-clamped inverters, IET Power Electron. 8 (4) (2015) 1–7.

[91] P.C. Loh, F. Blaabjerg, C.P. Wong, Comparative Evaluation of Pulse-Width Modulation Strategies for Z-Source Neutral-Point-Clamped Inverter, in: In Conference Proceeding in 37th IEEE Power Electronics Specialists Conference, 18–22 June, 2006.

[92] P.C. Loh, F. Blaabjerg, C.P. Wong, Comparative evaluation of Pulsewidth modulation strategies for Z-source neutral-point-clamped inverter, IEEE Trans. Power Electron. 22 (3) (2007) 1005–1013.

[93] P.C. Loh, D.G. Holmes, Y. Fukuta, T.A. Lipo, Reduced common-mode modulation strategies for cascaded multilevel inverters, IEEE Trans. on Industrial Applications 39 (5) (2003) 1386–1395.

[94] J.H.G. Muniz, E.R.C. da Silva, E.C. dos Santos Jr., A Hybrid PWM Strategy for Z-Source Neutral-Point-Clamped inverter, in: Conference Proceeding in APEC 2011, 2011, pp. 450–456.

[95] P.C. Loh, F. Gao, F. Blaabjerg, S.W. Lim, Operational analysis and modulation control of three-level Z-source inverters with enhanced output waveform quality, IEEE Trans. Power Electron. 24 (7) (2009) 1767–1775.

[96] F.B. Effah, P. Wheeler, J. Clare, A. Watson, Space-vector-modulated three-level inverters with a single Z-source network, IEEE Trans. Power Electron. 28 (6) (2013) 2806–2815.

[97] J. Pou, J. Zaragoza, S. Ceballos, M. Saeedifard, D. Boroyevich, A carrier-based PWM strategy with zero-sequence voltage injection for a three-level neutral-point-clamped converter, IEEE Trans. Power Electron. 27 (2) (2012) 642–651.

[98] Y. Liu, B. Ge, H. Abu-Rub, F.Z. Peng, Phase-shifted pulse-width-amplitude modulation for quasi-Z-source cascade multilevel inverter-based photovoltaic power system, IET Power Electron. 7 (6) (June 2014) 1444–1456.

[99] Y. Liu, B. Ge, H. Abu-Rub, F.Z. Peng, An effective control method for quasi-Z-source Cascade multilevel inverter-based grid-tie single-phase photovoltaic power system, IEEE Trans. Ind. Informatics 10 (1) (Feb. 2014) 399–407.

[100] Y. Liu, B. Ge, H. Abu-Rub, F.Z. Peng, An effective control method for three-phase quasi-Z-source cascaded multilevel inverter based grid-tie photovoltaic power system, IEEE Trans. Ind. Electron. 61 (12) (Dec. 2014) 6794–6802.

[101] Y. Liu, B. Ge, H. Abu-Rub, F.Z. Peng, A Modular Multilevel Space Vector Modulation for Photovoltaic Quasi-Z-Source Cascaded Multilevel Inverter, in: Conference proceeding in IEEE APEC 2013, 2013, pp. 714–718.

Index

Note: Page numbers followed by *f* indicate figures and *t* indicate tables.

A

Active-neutral-point-clamped (ANPC) inverter, 29, 233
Adjustable speed drives (ASDs), 2–3, 3*f*, 6
Alternative phase opposition and disposition (APOD), 23–24, 120, 158, 248–251, 250*f*
Asymmetric MLI (A-MLI) topologies, 181–182, 182*f*, 212*f*
   binary and trinary voltage output levels of, 183*f*
   with polarity generation part, 184–197
      multilevel DC link inverter (ML-DCLI), 184–189
      reduced component asymmetric multilevel inverter (RCA-MLI), 195–197, 196*f*, 197*t*
      simplified asymmetric multilevel inverter (SA-MLI), 189–192, 190–192*f*
      switched capacitor cell hybrid multilevel inverter (SCH-MLI), 193–195, 193–194*f*, 195*t*
   without polarity generation module, 197–210
      asymmetric cascade multilevel inverter (AC-MLI), 197–200, 198*f*, 199*t*, 201*t*
      cascaded basic blocks multilevel inverter (CBB-MLI), 200–203, 201*f*, 201*t*, 203*f*, 204*t*
      cascaded modified H-bridge multilevel inverter (CMB-MLI), 203–204, 205*f*
      cross connected sources based multilevel inverter (CCS-MLI), 204–210, 206–207*f*, 207–208*t*
Auxiliary resonant commutated poles (ARCPs), 237

B

Bipolar SPWM, 63–65, 65–66*f*

C

Cascaded basic blocks multilevel inverter (CBB-MLI), 200–203, 201*f*, 201*t*, 203*f*, 204*t*
Cascaded half-bridge (CHB) SM, 156, 156*t*
Cascaded H-bridge (CHB) multilevel topology, 3–5, 11–13, 14*f*, 141–142, 234
Cascaded H-bridge converter, 148
Cascaded modified H-bridge multilevel inverter (CMB-MLI), 203–204, 205*f*
Circulating current control, 167–170
Class D full-bridge LCL resonant inverter, 20*f*
Class D resonant inverter topologies, 18, 19*f*
Common mode voltage (CMV) and leakage current, 60–63
Continuous input current (CIC), 223–224
Continuous modified reference (MR) PWM, 248
Controlled switch network (CSN), 218–219
Control schemes, fundamentals of, 23–26
Conventional multilevel inverter topologies, 6–15
   flying capacitor, 10–11
   H-bridge MLI, 11–15
   neutral point clamped (NPC), 8–10
Cross connected sources based multilevel inverter (CCS-MLI), 204–210, 206–207*f*, 207–208*t*
   modular parallel sources multilevel inverter (MPS-MLI), 196*f*, 208–210, 209–210*f*, 210*t*
Current source converters (CSCs), 170–171
Current source inverter (CSI), 1, 5, 23, 223

D

DC-AC converter devices, 1
DC-link cascaded (DCLC) inverter, 242
Diode clamped converter (DCC), 148
Diode clamped MLI topology. *See* Neutral point clamped (NPC)
Direct torque control (DTC) method, 173–174
Discontinuous conduction mode (DCM), 239
Discontinuous input current (DIC), 223
Distribution STATCOM (D-STATCOM) active harmonic filters (AHFs), 1–2
Dynamic voltage restorer (DVR), 137–138, 174

E

Edge insertion (EI) PWM, 248
Electromagnetic interference (EMI) noise, 1–2

F

Field-oriented control (FOC) approach, 149, 159–160, 173–174
Finite set model predictive control, 116–118
Flexible AC transmission systems (FACTS), 1–2

259

Index

Flying capacitor (FC) topologies, 3–5, 3*f*, 10–11, 11*f*, 148, 233
Full-bridge (FB) inverters, 242. *See also* H-bridge inverter topology
Full-bridge (FB) SM, 154–155
Full-bridge LLC resonant inverter, 21

G

Galvanic isolation, 57–58
Grid-connected PUC7 inverter, 112–113, 116–117, 117*f*, 118*t*
Grid interactive inverters, 1–2

H

H-bridge inverter topology, 59–60
  common mode voltage (CMV) and leakage current, 60–63
  H4 inverter topology, 63, 64*f*
  H5 inverter topology, 70–72, 74*t*, 74*f*, 76–77*f*, 101*t*
  H6 inverter topology, 72–77, 78–80*f*, 80*t*
  highly efficient and reliable inverter concept (HERIC) inverter, 77–81, 82*f*, 83*t*, 84–85*f*, 105–106*f*, 106*t*
  modulation strategy, 63–70
    bipolar SPWM, 63–65, 65–66*f*
    hybrid SPWM, 68–70, 71*f*, 72*t*
    unipolar SPWM, 66–68, 67*f*, 69*f*, 70*t*
  recent H-bridge based multilevel topologies, 81–104
    active clamped HERIC topology, 99–104
    H6-I inverter topology, 83–87, 89*f*, 91*t*
    H6-II inverter topology, 83–87, 88*f*, 91*t*
    H6-III inverter topology, 88, 92–94*f*, 94*t*
    H6-IV inverter topology, 88–92, 94–96*f*, 96*t*
    H-bridge zero voltage rectifier (HB-ZVR) topology, 98–99, 99*f*, 101*f*, 101*t*
    H-bridge zero voltage rectifier-diode type topology (HBZVR-D) topology, 99, 101–103*f*, 103*t*
    optimized H5 (oH5) topology, 83
    passive clamped H6 topology, 92–97
H-bridge MLI, 11–15
H-bridge zero voltage rectifier (HB-ZVR) topology, 98–99, 99*f*, 101*f*, 101*t*
H-bridge zero voltage rectifier-diode type topology (HBZVR-D) topology, 99, 101–103*f*, 103*t*
High-frequency (HF) transformer, 21–23
Highly efficient and reliable inverter concept (HERIC) inverter, 77–81, 82*f*, 83*t*, 84–85*f*, 105–106*f*, 106*t*

High-voltage alternating current (HVAC) transmission systems, 170–171
High-voltage direct current (HVDC) transmission systems, 148–149, 170–172
Hybrid SPWM, 68–70, 71*f*, 72*t*

I

Impedance-source (IS) networks, 223
  general operating principle of, 223–232
Impedance source-derived multilevel inverters, 239–251
Inductance quality, 17
In-phase disposition (IPD), 120
Insulated-gate bipolar transistors (IGBTs), 2, 43
Inverter. *See* DC-AC converter devices

L

LCCT networks, 224
Leakage current, 60–63
Leg voltage control, 161–162
Level-shifted PWM, 120
Load quality factor at resonance frequency, 21
Low-pass filter (LPF), 218
Lyapunov-based model predictive control, 124–126
Lyapunov function, 126

M

Maximum boost control (MBC), 228–229
Maximum constant boost control (MCBC), 228–229
Medium-voltage motor drives, 148–149, 172–174
Metal oxide semiconductor field effect transistor (MOSFET), 2
Model predictive control (MPC), 35–36, 38–40, 115–116, 124–125
Modified reference (MR) PWM, 248
Modified space vector modulation control (MSVMBC), 228–229
Modular multilevel converter (MMC), 148–149
  applications, 170–174
    high-voltage direct current (HVDC) transmission systems, 170–172
    medium-voltage motor drives, 172–174
    offshore wind farms, 172
    power quality improvement, 174
  centralized control method for, 157*f*
  classical control methods, 156–158
  control objectives for, 157*f*
  current control, 165–170
    circulating current control, 167–170
    output current control, 165–167

principle of operation, 151–152
pulse width modulation (PWM) schemes, 158–161, 159f
  phase-shifted carrier modulation scheme, 159–160
  staircase modulation scheme, 160–161
submodule capacitor voltage control, 161–164
  leg voltage control, 161–162
  voltage balancing strategy, 162–164
submodule configurations, 152–156
Modular parallel sources multilevel inverter (MPS-MLI), 208–210
Modulation techniques, 33–35
  sinusoidal pulse width modulation (SPWM), 34–35
  space vector pulse width modulation (SV-PWM), 35
Multicarrier pulse width modulation, 118–124
  experimental results and discussion, 122–124
  level-shifted PWM, 120
  phase-shifted PWM, 120–122
Multilevel DC link inverter (ML-DCLI), 184–189, 187f
Multilevel inverters (MLIs), 1–6, 3–4f, 181, 232–235
  classification of, 233f
  impedance source-derived, 239–251
  modulation methods, 6f
  resonant, 236–238
Multilevel voltage source inverters
  control schemes, fundamentals of, 23–26, 24f
  conventional multilevel inverter topologies, 6–15
    flying capacitor, 10–11
    H-bridge MLI, 11–15
    neutral point clamped (NPC), 8–10
  soft switching and resonant multilevel inverters, 15–23
Multiresonant buck converter, 219f

N

Nearest three vectors (NTVs) modulation principle, 246–248
Nested neutral-point clamped (NNPC) converter, 148
Neutral point clamped (NPC) inverter topology, 2–5, 3f, 8–10, 10f, 29, 37t
Neutral-point-clamped multilevel inverters, 29–43
  converter configuration, 30
  modulation techniques, 33–35
    sinusoidal pulse width modulation (SPWM), 34–35
    space vector pulse width modulation (SV-PWM), 35
  switching states and commutation, 31–33
  three-level neutral-point-clamped inverter configuration, 30f
  three-phase three-level neutral-point-clamped inverter, 35–43
    model predictive current (MPC) control, 38–40
    simulation results, 40–43
    system model, 37–38
Newton-Raphson iterations, 25–26

O

Offshore wind farms, 148–149, 172
Optimized H5 (oH5) topology, 83, 85–87f
Output current control, 165–167

P

Packed U-cell (PUC) topology, 5, 112–115, 143–144
  applications, 135–140
    grid-connected mode, 136
    PUC5-based DVR, 137–138
    PUC5-based STATCOM, 137
    PUC5 rectifier, 136
    PUC5 three-phase inverter, 138–140
    PUC7 grid-connected inverter, 112–113, 116–117, 117f
    stand-alone mode, 136
  commercialization challenges, 141–143
    achieving product/market requirement, 142–143
    building a mass-producible product out of a laboratory concept, 142
    keeping the costs/benefits/reliability balance over time, 143
  control challenges, 115
  finite set model predictive control, 116–118
  Lyapunov-based model predictive control, 124–126
  mathematical modeling, 113–114
  multicarrier pulse width modulation, 118–124
    experimental results and discussion, 122–124
    level-shifted PWM, 120
    phase-shifted PWM, 120–122
  reduced sensor control, 128–135
  sliding mode control, 127–128
Parallel-resonant circuit (PRC), 220
Passive clamped H6 topology, 92–97
Phase disposition (PD), 23–24
  PD PWM scheme, 158

Phase-opposition disposition (POD) PWM scheme, 23–24, 120, 158
Phase-shifted carrier modulation scheme, 159–160
Phase-shifted PWM, 120–122
Power quality improvement, 174
Power semiconductors, switching methods applied to, 16–17, 16f
Pulse width modulation (PWM), 2, 23–24, 48–49, 150, 158–161, 228, 246–248
    phase-shifted carrier modulation scheme, 159–160
    staircase modulation scheme, 160–161

Q

Quasi-impedance source (qZS) inverter, 23, 223
Quasi-resonant source inverters (qZSI), 5

R

Reduced common-mode (RCM) PWM, 248, 249f
Reduced component asymmetric multilevel inverter (RCA-MLI), 195–197, 196f, 197t
Reduced element count (REC), 251
Reduced sensor control, 128–135
Renewable energy sources (RESs), 57, 112, 181
Resonant and Z-source multilevel inverters
    impedance source-derived multilevel inverters, 239–251
    impedance-source networks, general operating principle of, 223–232
    resonant circuits, general operating principle of, 218–222
    resonant multilevel inverters, 236–238
Resonant circuits, general operating principle of, 218–222
Resonant converters, classification of, 218t
Resonant microinverter applications, 22f
Resonant multilevel inverters, 236–238
Resonant power converters, 21–23
Resonant tank network (RTN), 218
Resonant-transition converter, 219f
Root Mean Square (RMS) value, 149–150

S

Sampled average modulation (SAM) scheme, 158–159
Selective harmonic elimination (SHE), 159
Selective harmonic elimination PWM (SHE-PWM), 5–6, 24
Series-resonant circuit (SRC), 220
Series resonant network, frequency spectrum of, 17f
Seven-level boost active-neutral-point-clamped (7 L-BANPC) circuits, 29
Shoot-through (ST) cross-conduction, 223
Simple boost control (SBC), 228–229
Simplified asymmetric multilevel inverter (SA-MLI), 189–192, 190–192f
Single-DC-source multilevel inverter (SDCS-MLI) topology, 112
Single-phase NPC inverter, switching states of, 31t
Single-supply inverters, 3–5
Sinusoidal pulse width modulation (SPWM), 5–6, 23–24, 34–35, 58–59
    bipolar SPWM, 63–65, 65–66f
    hybrid SPWM, 68–70, 71f, 72t
    unipolar SPWM, 66–68, 67f, 69f, 70t
Sliding mode control, 127–128
Soft switching and resonant multilevel inverters, 15–23
Soft-switching methods, 16–17
Space vector modulation (SVM) scheme, 158–159
Space vector pulse width modulation (SV-PWM), 5–6, 26, 35
Stacked multicell (SMC) converter, 148
Staircase modulation scheme, 159–161
Static synchronous compensator (STATCOM), 148–149, 174
Submodule capacitor voltage control, 161–164
    leg voltage control, 161–162
    voltage balancing strategy, 162–164
Switched capacitor cell hybrid multilevel inverter (SCH-MLI), 193–195, 193–194f, 195t
Switching power loss, 16–17
Switching states and commutation, 31–33
Switch-open circuit (SOC) fault, 46–48
Switch short-circuit (SSC) fault, 46, 48

T

Third harmonic injection PWM (THI-PWM), 5–6
Three-level DCC, 148
Three-level inverter, 29
Three-phase three-level neutral-point-clamped inverter, 35–43
    model predictive current (MPC) control, 38–40
    simulation results, 40–43
    system model, 37–38
Total harmonic distortion (THD), 1–5, 29, 35, 43–44, 48–49, 60, 123–124
T-type inverter, 43–54
    DC capacitor voltages, switching states on, 50–51
    modulation of, 48–50
    operating principle, 43–45
    simulation results, 52–54
    switch open-circuit fault, 46–48

switch short-circuit fault, 48
T-type NPC topologies, 5

U
Unified power flow controllers (UPFCs), 1–2
Unified power quality compensator (UPQC), 174
Unipolar SPWM, 66–68, 67f, 69f, 70t

V
Variable frequency drives (VFDs), 2
Voltage balancing strategy, 162–164
Voltage-oriented control (VOC) approach, 149, 159–160
Voltage source converters (VSCs), 170–171
Voltage source inverter (VSI), 1, 5, 7f, 7t, 9f, 13t, 23, 29, 223

W
Wind farms (WFs), 172
  offshore, 148–149, 172

Z
Zero current switching (ZCS) method, 16–18, 218
Zero transition switching, 15f
Zero voltage switching (ZVS) method, 16–18, 218, 236
Z-source inverter (ZSI), 223–224